Interactive Approaches to Water Governance in Asia

Kenji Otsuka
Editor

Interactive Approaches to Water Governance in Asia

Springer

Editor
Kenji Otsuka
Institute of Developing Economies
Japan External Trade Organization (IDE-JETRO)
Chiba, Japan

ISBN 978-981-13-2398-0 ISBN 978-981-13-2399-7 (eBook)
https://doi.org/10.1007/978-981-13-2399-7

Library of Congress Control Number: 2018963999

This Springer imprint is published by the registered company Springer Nature Singapore Pte Ltd.
The registered company address is: 152 Beach Road, #21-01/04 Gateway East, Singapore 189721, Singapore

Preface

This book is a major result of the 2-year research project on "Interactive Approaches to Water Governance: Case Studies in Asia" conducted at the Institute of Developing Economies, Japan External Trade Organization (IDE-JETRO), from the fiscal year of 2015–2016 to 2016–2017.

This book introduces interactive perspectives, which have been discussed mainly in the context of Western European countries, to case studies on water governance in Asia. It examines how these perspectives reveal complex and dynamic interactions in water governance in Asia and how interactions between policies and practices as well as interactions between formal institutes and emerging informal institutes occurred. Through our case studies in Asia (from Japan, China, Thailand, the Philippines, Indonesia, and India), the book reveals that there are emerging interactive forms in Asia under hierarchical but fragmented administrative system, although water resource and environmental management is still occupied with governmental sectors. Also, it should be noted that hybrid forms of interactive governance including governmental and non-governmental actors have been emerging and under such forms the expected role of government as well as non-governmental actors could be changed in more cooperative way to solve the problems. In such forms, researchers outside the locality could play an important role to facilitate such a form of interactive governance.

The feature of interactive forms and the role of such a transdisciplinary approach can be learned through this book by not only scholars and university students, but also policy-makers and grassroot practitioners involved themselves in interactive process of water governance in and beyond the region. In terms of academic disciplines, this book can cover water and environmental governance, water and environmental resource management, public policy, political science, sociology, and other related social sciences focusing on water and environmental issues, and area studies in Asia generally.

This book could not be completed without generous support and advice from numerous scholars and practitioners in our field research, research meetings, and writing papers. Also, we would like to express our appreciation to all staff at IDE-JETRO in managing our research project and all staff at Springer Nature for editing this book.

Chiba, Japan Kenji Otsuka

Contents

Chapter 1
Interactive Perspectives on Water Governance in Asia

Kenji Otsuka

Abstract This chapter argues for the importance of water governance perspectives in addressing water issues in Asia, focusing on interactions between policy and practice and between formal institutions and informal practices. First, the concepts of IWRM are revisited by referring to international guidelines. Next, major international perspectives on water governance as well as the concept of basin governance are reviewed, followed by ways to develop proper concepts for Asian countries and communities through interactive approaches. Finally, this chapter identifies issues to be discussed in each chapter of this book and synthesizes our findings while addressing remained tasks to be pursued further.

Keywords Water governance · Integrated water resource management (IWRM) · Basin governance · Interactive governance · Interactive approaches · Asia

1 Introduction: Seeking Approaches to Water Governance in Asia

For several decades, the international community has been calling for effective responses to urgent water crises. The 1977 United Nations Conference on Water at Mar del Plata, Argentina, was the first UN conference focusing on water. Since the 1990s, more than a dozen international conferences and numerous global agenda statements have addressed water concerns. The first World Water Development Report (WWDR), *Water for People, Water for Life*, was published in 2003, in conjunction with the Third World Water Forum in Kyoto, Japan. Furthermore, UN-Water, the UN interagency coordination mechanism for all issues related to freshwater, was also established in 2003. Since 2003, the WWDR has been published triennially, providing informative, scientific, and policy-oriented knowledge

K. Otsuka (✉)
Institute of Developing Economies, Japan External Trade Organization (IDE-JETRO), Chiba, Japan
e-mail: kenji_otsuka@ide.go.jp

K. Otsuka (ed.), *Interactive Approaches to Water Governance in Asia*,
https://doi.org/10.1007/978-981-13-2399-7_1

on water in more than a thousand pages of text across its five editions so far (WWAP 2003, 2006, 2009, 2012, 2015).

Over the years, a significant number of water reports have been published, indicating that our world is facing persistent present and projected water crises in numerous forms and contexts, as well as over a long-term span. Water, as we all know, is essential not only for human life but also for other species in our planet's ecosystem (WWAP 2003). Along with supporting life, water also has the potential of taking it away,[1] through overabundance, scarcity, or over-pollution of water sources. Furthermore, such conditions vary and fluctuate across time, space, and social dimensions (WWAP 2012).

In East Asia and South Asia, where our project is focused, there is a broad array of water crises and stresses, such as water scarcity, land subsidence due to overextraction of groundwater, poor access to safe drinking water and to improved sanitation, eutrophication of lakes and other waters, droughts and floods resulting from extreme weather conditions, conflicts in water infrastructure development, and threat of rising sea levels along coastal areas, caused by global warming (WWAP 2012, 2015). These challenges occur in varied climates and complex geographies and are further complicated by high population density, rapid industrialization and urbanization, and increasing social inequalities in the region.

In addressing unsolved and projected water issues in this region and beyond, the international water community and its member states have been developing and disseminating various agendas, guidelines, action plans, information kits, tool boxes, and other forms of information and knowledge all over the world. Among them, integrated water resource management (IWRM) is a popular concept accepted by the global water community and by several individual countries and regions (GWP-TEC 2000).

IWRM and similar concepts developed from it, such as integrated river basin management (IRBM) and integrated lake basin management (ILBM), are useful but not a panacea. Water crises and stresses are rooted not only in water's physical availability but also in issues of "power, poverty, and inequality"; "the underlying cause of scarcity in the large majority of cases is institutional and political, not a physical deficiency of supplies" (UNDP 2006).

Solutions to water issues require action not only from within the "water box"—that is, the specialized realm of water resource and water environmental management—but by people from many other fields. "Many paths to sustainable development are linked to water, but the decisions that determine how water resources are used or abused are not made by water managers alone" (WWAP 2009, preface). Socioeconomic decision-making through interactions between governments and other political actors, businesses and other economic actors, and civil society actors outside the "water box" influence water management and vice versa. For this reason, to solve water issues, we must address questions of water governance.

[1] Editors' Introduction, Radical History Review, 116, 2013, 1.

Water management and related actions, even if wisely conceived, will fail if not accompanied by sustainable local practices at any given site. Therefore, both policies and practices should always matter in water governance, especially in the implementation and monitoring of water agendas at every level.

This article argues for the importance of water governance perspectives in addressing water issues in Asia, focusing on interactions between policy and practice and between formal institutions and informal practices. In the next section, we revisit the concepts of IWRM by referring to international guidelines. Next, we review international perspectives on water governance and also address the concept of basin governance, followed by ways to develop proper concepts for Asian countries and communities through interactive approaches. Finally, we identify issues to be discussed in each chapter of this book and synthesize our findings while addressing remained tasks to be pursued further.

2 IWRM Revisited

The concept of IWRM was developed through a series of international consultations on water issues, such as the 1977 UN Conference on Water at Mar del Plata, the International Conference on Water and Environment in Dublin and the UN Conference on Environment and Development (UNCED) in Rio de Janeiro in 1992, the World Water Forum held triennially since 1997, and others (Pangare et al. 2006). These international discussions have sought to establish holistic and comprehensive approaches to water issues, encompassing life sustenance for people and ecosystems, resources for socioeconomic development, prevention of hazards, reduction of risks, and participation by all stakeholders. The existence of monopolistic sectoral approaches to water resource management, which lead to "fragmented and uncoordinated development and management," has led to broad recognition of a "water governance crisis" (GWP-TAC 2000, 9). The Dublin principles, incorporated into Agenda 21 at UNCED,[2] are said to provide the basis for the international understanding of IWRM. These principles include four pillars: (1) freshwater is a finite and vulnerable resource, essential to sustain life, development, and the environment; (2) water development and management should be based on a participatory approach, involving users, planners, and policy-makers at all levels; (3) women play a central part in the provision, management, and safeguarding of water; and (4) water has an economic value in all its competing uses and should be recognized as an economic good (GWP-TAC 2000, 13–21).

On the basis of this international agreement to address water issues, the Global Water Partnership (GWP) was established in 1996 to foster IWRM. It defines IWRM as "a process which promotes the co-ordinated development and management of water, land and related resources, in order to maximize the resultant

[2] The principles are also referred to as the '"Dublin-Rio principles" (GWP-TAC 2000, 13).

economic and social welfare in an equitable manner without compromising the sustainability of vital ecosystems" (GWP-TAC 2000, 22).

As implied by this definition, IWRM has "overriding criteria" in economic, social, and environmental (ecological) dimensions. In the economic dimension, it requires "economic efficiency in water use," which means that "water must be used with maximum possible efficiency" in view of "the increasing scarcity of water and financial resources, the finite and vulnerable nature of water as a resource, and the increasing demands upon it." In the social dimension, it requires "equity," which means that "the basic right for all people to have access to water of adequate quantity and quality for the sustenance of human wellbeing must be universally recognized." And in the environmental and ecological dimension, it requires "environmental and ecological sustainability," which means ensuring that how water resources are used "does not undermine the life-support system thereby compromising use by future generations of the same resource" (GWP-TAC 2000, 30).

The GWP also identifies essential principles with regard to integration, which should be considered both within and between natural and human (social) systems across time and space. Recommended integrations of natural systems include freshwater management and coastal zone management, land and water management, "green water" and "blue water", surface water and groundwater management, quantity and quality, and upstream and downstream interests. As for human systems, integration involves mainstreaming of water resources, cross-sectoral integration in national policy development (including both clean water and wastewater management), macroeconomic effects of water developments, integrated policy-making, influence on decisions made in the economic sector, and participation by all stakeholders (GWP-TAC 2000, 23–29). Every integration process should involve a careful management of interactions between both systems.

Such integrations are needed to overcome the monopolistic, sectoral, and fragmented forms of water resource management that cause water governance crises; however, "integration per se cannot guarantee development of optimal strategies, plans and management schemes" (GWP-TAC 2000, 23). Moreover, "there is no universal blueprint as to how such principles can be put into practice," and "IWRM practices depend on context," for instance, "the nature, character and intensity of water problems, human resources, institutional capacities, the relative strengths and characteristics of the public and private sectors, the cultural setting, natural conditions and many other factors" that "differ greatly between countries and regions" (GWP-TAC 2000, 6–7).

During the two decades since the GWP was established, considerable global experience in IWRM implementation has been accumulated. Some of these experiences have been archived on the "IWRM Toolbox" page of the GWP website, which includes not only national-level but also basin- and local-level cases. The site includes 48 cases from Africa, 47 from the Americas and the Caribbean, 64 from Asia and the Caucasus, 7 from Australia and Oceania, 60 from Europe, and 10 from the Mediterranean and the Middle East, for a total of 236 cases.[3] They include

[3]Accessed on 29 January 2016.

success stories as well as ongoing and unsuccessful cases offering lessons to be learned. Additionally, the Toolbox provides 59 different instruments for implementing IWRM arranged in three categories: (1) the enabling environment, (2) institutional roles, and (3) management instruments (see Annex 1.1). A user can identify actual cases in which each instrument was used to help in determining whether a given strategy would work well in a particular context.

In addition to case studies and instruments, the IWRM Toolbox also describes "critical challenges" and "cross-cutting issues." IWRM critical challenges are water and related nexus issues such as climate change, food security, urbanization, energy security, and ecosystems. The cross-cutting issues involve multiple domains of social, economic, and environmental development; they include gender, youth, financing, governance, and cooperation.[4] As the Toolbox indicates, however, "no blueprint for the application of IWRM can be given," and therefore, the IWRM implementation task requires us to "select a suitable mix and sequence of processes or steps that work in a given situation, context and country." In addition to this intrinsic characteristic of IWRM, case studies described on the website provide only summaries and outlines, not a full range of experiences and lessons to be applied when we face practical problems in the field.

Thus, to overcome the gap between policies and practices in water governance, we must look into the process of water governance more carefully through case studies, focusing on some key aspects to construct an operational framework without neglecting the process's complexities.

3 International Perspectives on Water Governance

3.1 Effective Water Governance

To put IWRM into practice effectively, we must return to the concept of water governance because failure in this area impedes comprehensive IWRM implementation. The GWP defines water governance as follows:

> Water governance refers to the range of political, social, economic, and administrative systems that are in place to develop and manage water resources, and the delivery of water services, at different levels of society. (GWP 2003, 7, 16)

In addition, water governance "embraces the formal and informal institutions by which authority is exercised" (GWP 2003, 7). Thus, water governance is considered to encompass broad perspectives in political, social, economic, and administrative dimensions, as well as both formal and informal institutions participating in water resource development and management. Importantly, water governance focuses not only on "internal governance" (traditional water management) within the water sector (the "water box" mentioned earlier) that often emphasizes economic and

[4] See the Toolbox page of the GWP website at http://www.gwp.org/en/ToolBox/.

technical solutions but also on "external governance" outside the water sector, which involves processes more political in nature (GWP 2003, 17).

These perspectives indicate that water policy should be implemented effectively, and the key actors (stakeholders) should be involved in its formation as well as its implementation. "Effective water governance" means governance that permits effective formation and implementation of water policy and enables application of IWRM to develop, allocate, and manage water use "equitably and efficiently and ensuring environmental sustainability" while requiring that "disparate voices are heard and respected in decisions over common waters and use of scarce financial and human resources" (GWP 2003, 16).

Although there is "no single mode of effective governance," there do exist "some basic principles or attributes that are considered essential." The GWP identifies seven principles of effective water governance in two categories, "approaches" and "performance and operation," as described below (GWP 2003, 26–29):

Approaches

1. Open and transparent: All policies are open and transparent so that both insiders and outsiders can easily follow the steps taken in the policy formation.
2. Inclusive and communicative: Participation crucially depends on all levels of government following an inclusive approach when developing and implementing policies. Governance institutions and systems need to communicate between the actors and stakeholders in very direct ways.
3. Coherent and integrative: Policy and action must be coherent. Coherence requires political leadership and a strong responsibility on the part of the institutions at different levels to ensure a consistent approach within a complex system. Water governance should enhance the effectiveness of IWRM.
4. Equitable and ethical: Equity between and among the various interest groups, stakeholders, and consumer-voters needs to be carefully monitored throughout the process of policy development and implementation. Water governance has to be strongly based upon the ethical principles of the society in which it functions and based on the rule of law.

Performance and Operation

5. Accountable: Decision-makers in government, the private sector, and civil society organizations are accountable to the public, as well as to institutional stakeholders.
6. Efficient: In addition to economic efficiency, there need to be concepts of political, social, and environmental efficiency to balance between different dimensions of efficiency. Minimizing transaction costs will go a long way toward political and economic efficiency.
7. Responsive and sustainable: Responsiveness requires policies to be implemented in a proportionate manner and decisions to be taken at the most appropriate level. The institutions should be built with an eye toward long-term sustainability.

The previous statement of effective GWP water governance seems to present a "one-size-fits-all" concept, but it also reminds us of the difficulties in applying the same IWRM strategy to developing as to developed countries, since developing countries frequently have poor capacity in administration, democratic institutions, social service provision, and other factors. In terms of governance system reform, the GWP indicates that "developed countries [are] moving towards flexibility and distributed governance systems whilst developing countries are characterized by rigidity and hierarchical and light governance systems" (GWP 2003, 34).

The GWP's discourse on effective water governance is helpful in reminding us to pay more attention to the broader aspects of water issues, including social and political factors; however, it is implied that the development and application of IWRM tools can overcome governance failures and crises. These are conflicting concepts because it seems that IWRM implementation requires effective water governance, but developing and introducing IWRM tools can address the failure of governance. In other words, the discourse on effective water governance emphasizes primarily functional solutions using IWRM tools although that discourse does present rich perspectives on water governance.

3.2 OECD Principles of Water Governance

The Organisation for Economic Co-operation and Development (OECD) is also concerned with water crises and recognizes that they "are often primarily 'governance' crises" caused by fragmentation "across sectors, places and people, as well as geographic and temporal scales." To provide an operational framework to "consider the short, medium and long term in a consistent and sustainable way" for IWRM application, water governance principles are drafted through discussions among member states as well as multi-stakeholder participants including public, private, and nonprofit sectors (OECD 2015, 1–7). The OECD lists 12 principles in 3 main categories (OECD 2015, 9–12):

Effectiveness Principles

1. Clear roles and responsibilities: Clearly allocate and distinguish roles and responsibilities for water policy-making, policy implementation, operational management, and regulation, and foster coordination across these responsible authorities.
2. Appropriate scales within basin systems: Manage water at the appropriate scale(s) within integrated basin governance systems to reflect local conditions, and foster coordination between the different scales.
3. Policy coherence: Encourage policy coherence through effective cross-sectoral coordination, especially between policies for water and the environment, health, energy, agriculture, industry, spatial planning, and land use.
4. Capacity: Adapt the level of capacity of responsible authorities to the complexity of the water challenges to be met and to the set of competencies required to carry out their duties.

Efficiency Principles

5. Data and information: Produce, update, and share timely, consistent, comparable, and policy-relevant water and water-related data and information, and use it to guide, assess, and improve water policy.
6. Financing: Ensure that governance arrangements help mobilize water finance and allocate financial resources in an efficient, transparent, and timely manner.
7. Regulatory frameworks: Ensure that sound water management regulatory frameworks are effectively implemented and enforced in pursuit of the public interest.
8. Innovative governance: Promote the adoption and implementation of innovative water governance practices across responsible authorities, levels of government, and relevant stakeholders.

Trust and Engagement Principles

9. Integrity and transparency: Mainstream integrity and transparency practices across water policies, water institutions, and water governance frameworks for greater accountability and trust in decision-making.
10. Stakeholder engagement: Promote stakeholder engagement for informed and outcome-oriented contributions to water policy design and implementation.
11. Trade-offs: Encourage water governance frameworks that help manage trade-offs across water users, rural and urban areas, and generations.
12. Monitoring and evaluation: Promote regular monitoring and evaluation of water policy and governance where appropriate, share the results with the public, and make adjustments when needed.

These principles are developed "on the premise that there is no one-size-fits-all solution to water challenge worldwide," that "water policies need to be tailored to different water resources and places, and that governance responses have to adapt to changing circumstances" (OECD 2015, 5). They also resemble the GWP principles of effective water governance, with IWRM, in their reliance on functionalism to provide practical, operational, and pragmatic tools for better water governance. Within their perspective on effective water governance, OECD principles give us new viewpoints, such as "innovative governance" as an efficiency principle, as well as "trade-offs" across users, rural and urban areas, and generations as a principle of trust and engagement. These can give us rich insights for effective water governance. However, they seem to focus very little on equity or on disaster risk reduction for affected communities' vulnerability and resilience. These factors of equity, vulnerability, and resilience are all more important in developing than in developed countries.

4 Basin Governance Perspectives

To put IWRM into practice, the basin (watershed) scale must receive primary focus. When we examine geographical boundaries in regions all over the world, we find mismatches between administrative jurisdictions and watersheds in many countries and along international rivers. As for regional water governance, the basin should be an important unit for integrating water resources and other resources such as land, fields, and forests, which are usually managed and monitored by fragmented administrations. The basin should also be a key unit for integrating ecological, social, cultural, and political systems, which are understood and studied by different disciplines. Pursuant to this viewpoint, "river basin governance" (Turner and Otsuka 2005) and "lake basin governance" (RCSE-SU and ILEC 2014) [hereafter described collectively as "basin governance"] have been extensively discussed in water policy studies and applied in actual practice.[5] In terms of integrated perspective, basin governance should extend "beyond water centricity" because "water governance is strongly interrelated with the aims and actions in a broader system of governance and action" (Teisman et al. 2013, 5).

First, basin governance emphasizes ecosystems' complexity and contingency in rivers, lakes, and their basins. According to the UN Environment Program's "Millennium Ecosystem Assessment" (MEA), "there is established but incomplete evidence that changes being made in ecosystems are increasing the likelihood of nonlinear changes in ecosystems (including accelerating, abrupt, and potentially irreversible changes), with important consequences for human well-being" (MEA 2005, 11). As for water problems, eutrophication and hypoxia in lakes and bays represent such instances. For example, during the last decade, lakes in China have often experienced proliferation of blue-green algal blooms under extreme weather conditions, such as high temperatures, without rainfall for a certain period. These algal blooms have resulted in water crises in the lake basin's coastal cities, where industrial, domestic, and agricultural wastewater have been dumped under rapid industrialization and urbanization. With regard to this and other environmental problems, "while science can often warn of increased risks of change it cannot predict the thresholds at which the change will be encountered" (MEA 2005, 11). In addition, once the threshold has been passed, considerable and long-lasting endeavors will be needed to reverse such an ecosystem transformation to normal (ILEC 2005, 13). This is why an adaptive approach to ecosystem management in water governance is necessary.

Second, basin governance focuses on institutional building, arrangement, and reform to make integrated basin management effective and sustainable through political, economic, and social processes. In particular, institutional reform is needed to overcome fragmented management of water and related resources in basins.

For example, during the past decade, Japan has made some progress in institution building and reform in water and basin policy at both central and local government levels. One such recent development is the Water Cycle Act and its Basic Plan. Historically, water resource and river management in Japan have been fragmented among multiple sectors of the central government. However, under the Water Cycle

[5]As a concept similar to "basin governance," we can also use "watershed governance." However, these two concepts differ somewhat because basin governance stresses upon geographical unity, and watershed governance places more stress upon functional integration of resources as water centrality. However, we know that "watershed" is sometimes used as a near-synonym for "basin."

Act, in effect since 2014, Japan is now moving toward multi-stakeholder cooperation in basins. Following this Act, in 2015, the Japanese Cabinet approved a Water Cycle Basic Plan that promotes cooperation in integrated basin management to regain and sustain sound water cycles nationwide. One action prescribed in the plan is to organize Basin Water Cycle Councils in each major river basin, to include a variety of stakeholders such as government river managers, local governments, industries, offices, farmers, residents, nonprofit organizations, and other tasks. Organizing these councils and enabling them to function effectively is a key challenge for basin governance in Japan.

Additionally, in basin governance, institutional building and reform cannot happen by themselves but must occur in conjunction with other tools and measures. Surveys of the world's lakes during the past decade have helped determine that governance for ILBM requires the following six pillars (RCSE-SU and ILEC 2014, 17):

1. Institutions to manage the lake and its basin for the benefit of all lake basin resource users
2. Policies to govern people's use of lake resources and their impacts on the lake
3. Involvement of people to facilitate all aspects of lake basin management
4. Understanding of technological possibilities and limitations, which often play a major role in dictating long-term decisions
5. Knowledge and information of traditional as well as modern scientific practices, forming the basis for informed decisions
6. Sustainable finance to support implementation of all activities listed above

Here, the main challenge is how these pillars can be well integrated and coordinated to create effective, sustainable basin governance. ILBM studies indicate that "basin stakeholders must achieve a greater level of inter-harmonization and inter-phasing of management reasons/purposes." Such inter-harmonization and inter-phasing could be enabled through "on-the-ground governance improvement," not necessarily at "higher level policy making at the national government level." This indicates that integration as a long-term process occurs by "necessity approach" as a unique process and not by design in a general setting. ILBM studies also suggest several different types of integration processes. First, integration by "encompassing" refers to "cross-sectoral coordination across government ministries and, for transboundary lakes, even different countries." Second, integration by "unification" provides a framework for promoting "a sense of mutual facilitation and collaboration" on the basis of "previously successful experiences, even if only marginal in some cases." Finally, integration by "evolution" involves expanding projects in "their spheres of operation over time, either spatially or sectorally." As an example of this last type of integration, a project initially focused on controlling point sources of pollution has expanded to controlling toxic contaminants, invasive species, and nonpoint sources of pollution in developing countries (RCSE-SU and ILEC 2014, 39–43).

In addition to these perspectives, another approach to enabling integrated management in basin governance can be found in a series of empirical field studies applying "research on the commons." As noted above, a basin has a variety of resources related to water, such as land, forest, and farm acreage. In numerous countries, these resources are not always managed by the central government or as private property but also by local communities or other types of local collectives (Ostrom 1990; Murota and Takeshita 2013). Studies applying discussions on "the commons" do not always target water and basin issues, but we can follow this model when considering possibilities and limits of "long-enduring, self-organized, and self-governed" organizations (Ostrom 1990, 58–61) for building basin governance institutions through a bottom-up approach.

For water and basin governance studies, focusing on the following points will be important: (1) how sustainable local institutions' components (including rules, methods of organization, and operating mechanisms) respond to internal and/or external changes or threats and (2) how local institutions can or should be (re)built or (re)formed to make basin governance effective and sustainable. As for the first item, there exist detailed case studies in Japan on communal land and forest management systems (Murota and Takeshita 2013). Synthesizing such case studies, one finds three types of responses by local institutions to external impacts such as population change, commoditization of common land, private or public investment in land development, and public or regulatory policy activities. These three types can be called "community-based fine-tuning adjustments," "collaborative adjustments," and "resistance to hostile external actors" (Mitsumata 2013). In addition, Otsuka (2015, 284–285), in field surveys covering both internal and external impacts on rural areas of East and Central Asian countries, identified an "innovative" response to ecological crisis situations. As one example, an ecological purification system for drinking water was developed to protect the health of villagers suffering from persistent pollution of rivers and groundwater in a Chinese river basin (Otsuka 2015, 260–262).

With regard to the second item, a unique social experiment involving community roundtable meetings took place in China for 4 years, beginning in 2008, conducted jointly by the Institute of Developing Economies (IDE), the Japan External Trade Organization, and the Center for Environmental Management and Policy (CEMP) of Nanjing University's School of the Environment (see Chap. 4 for more detail). As shown in Chap. 4, dialogue among local stakeholders is seen as promoting mutual trust and cooperation and supporting the establishment of informal, bottom-up institutions that can contribute to improved basin governance, although it has been difficult for these social sectors to organize themselves because the legitimacy of informal institutions is often uncertain under China's authoritarian system.

These discussions based on commons studies are very helpful, informing efforts in bottom-up institutional reform and establishment of basin governance; however, the nested structure of governance across local, regional (including basin-wide), national, and international scales remains a challenge for both research and actual practice. For instance, China does not merely have numerous administrative layers and stakeholders; the relation between central and local governments is complex, and many local governments are more concerned with pursuing pro-economic development policies than with ecological sustainability.

Besides seeking an effective institutional arrangement across different layers of government, one key concept in responding to this challenge is "analytic deliberation," which means "well-structured dialogue involving scientists, resource users, and the interested public, informed by analysis of key information about environment systems." Analytic deliberation should yield a solution that "provides improved information and the trust in it that is essential for information to be used effectively, builds social capital, and can allow for change and deal with inevitable conflicts well enough to produce consensus on governance rules" (Dietz et al. 2003).

5 Interactive Perspectives on Water Governance

These discussions on water and basin governance perspectives suggest that the core task is to discover how IWRM's required functional, institutional, and social integrations can be realized in adaptive, innovative ways beyond sectoral and fragmented water resource management, to ensure water centricity's suitable level in the basin management of complex, contingent social-ecological systems with unique local contexts[6] (Lubell and Edelenbos 2013; Teisman et al. 2013). We already have plenty of tools and instruments to understand, analyze, and improve IWRM practices, but we must consider wider contexts of political, economic, and social processes that enable or impede them, as discussed by institution-building perspectives on basin governance. Perspectives on institutional building through a bottom-up approach are still largely limited to some interactions between local communities and their outside world. We must move beyond this simple framework if we are to address complex, contingent water and basin governance issues consisting of multiple layers and many actors.

In this regard, it is worth noting that discussions about water governance have focused on interactions and connections across boundaries in problem-solving fields, which have been discussed in Western Europe (Edelenbos and Teisman 2013; Edelenbos et al. 2013; Teisman et al. 2013). The "water governance capacity" perspective in terms of connectivity identifies three types of boundaries to be considered: between functions of different policy areas, between levels of government, and between public and private domains. Governance capacity is understood as increasing when one is able to cross and span these boundaries (Edelenbos and Teisman 2013). Related to this ability, "the capacity to connect other domains, levels, scales, organizations and actors" is considered an important aspect of water governance (Edelenbos et al. 2013, 3). In this sense, "water governance is then about the ability to connect different frames, values and ambitions." This connective capacity is understood as "the capabilities of individuals, instruments and institutions to counter fragmentation in water governance processes by crossing boundaries (such as structure, organization, and language) and establishing linkages between different actors (on different levels, at various scales and in numerous domains) in

[6] For the concept of "social-ecological systems," see Berkes et al. (2003).

the light of solving water issues" (Edelenbos et al. 2013, 7). Although these concepts are developed in Western Europe, even in Asia, we can find similar cases if we are cautious of such aspects. For example, in Japan, environmental taxation on water and forest resources, introduced by prefectural governments in most of the country, have been considered a means of connecting different functions of forest management in the upper river basin and ecosystem services provided to urban residents, such as water supply, wood production, and recreation (Fujita 2005).

Boundary spanning across functions, levels, and domains in water governance is important for making connective capacity effective (Edelenbos and Teisman 2013; Edelenbos et al. 2013). Individuals can also play significant roles in this regard (Edelenbos et al. 2013, No. 4189). From this point of view, we can find some cases which tell us its importance in regional regeneration of underdeveloped regions in Asia. For example (although this case does not directly involve water issues), several mountainous rural communities in upper river basins in Japan have been threatened by rapid depopulation and aging, but some have been rejuvenated through community-based restaurants and shops operated by middle-aged residents who have returned to their hometowns after many years of urban living (Fujita 2015). In this case, the person returning from the city to his or her hometown could be considered a boundary spanner. In some cases, such as rural poverty alleviation projects in developing countries, not only individuals but also domestic and international nongovernmental organizations and also researchers from outside can become boundary spanners.

More generally, interactions across boundaries of functions, levels, and domains should receive greater attention in water governance studies. One recent study has highlighted the important function of "interactive governance in terms of quasi markets, partnerships, and governance networks that are a prominent part of the new and emerging reality and seem to both challenge and transform the role of government in governing society and the economy." This study further contends that "government, markets, interactive forms of governance, or some mixture of the three will offer the 'right' governance solution" (Torfing et al. 2012, 5, 11). Here, let us look into the background of interactive governance theory briefly to consider the validity of its application to case studies on water governance in Asia.

In the Netherlands, where we can find a lot of literatures on water governance studies in terms of interactive perspectives in Western Europe, as described in details at the next chapter, according to experiences of political and social change, local governments are "moving from a central steering paradigm to an interactive steering paradigm"; after "a historically low level in voter turnout at the local election in 1990, numerous initiatives for policy renewal have emerged in an attempt to diminish the (perceived) legitimacy gap between local politicians and the electorate at large" (Edelenbos 2005, 112). "Interactive governance is often organized as an informal process with particular rules and roles that are different from the existing institutional representative system and run parallel or prior to the formal institutions of negotiation and decision making." Therefore, interactive governance should be focused on "a missing institutional link between the interactive process and the formal decision-making process" (Edelenbos et al. 2010, 74).

In interactions between government, as a center of formal institutions, and the public, as stakeholders of informal institutions in many cases, "public participation" or "citizen participation" is a popular form in which the two entities revolve in environmental and resource governance in many parts of the world. This form of interactive governance is "government induced interactive governance." On the contrary, we can find another form called "citizen induced interactive governance" (Edelenbos and van Meerkerk 2016). In this regard, "self-organized, and self-governed organizations" (Ostrom 1990, 58–61) for building basin governance institutions through a bottom-up approach (as discussed above) could be one of these forms.

The theory of interactive governance has been developed under the transition in Western democratic system; however, we can find some key concepts worthwhile applying in the research field of water governance in Asia and other regions. As we see in previous sections, there is a consensus in international water policy communities that multi-stakeholder involvement cannot be avoided in the process of water governance not only in Western developed countries but also in non-Western developed and developing countries. So, the research question in case studies on water governance in Asia should be focused not on whether or how interactive governance could be applied but on what kinds of form could be found under interactions among stakeholders especially between the government, which is still considered as a center of water governance in many countries in Asia, and nongovernmental actors, which are considered to be emerging player with more important roles than before.

In terms of interactions between the government and nongovernmental actors in public policy process, the theory of interactive governance gives us important perspectives. One is the concept of "metagovernance" or "governance of governance." In this regard, government can play a critical role "by means of shaping its structural and institutional conditions and by designing, managing, and directing the interactive governance arenas" (Torfing et al. 2012, 4–5). Thus, "the notion of metagovernance offers a way of balancing state-centered and society-centered views on how society and the economy are governed" (Torfing et al. 2012, 132). In metagovernance initiatives, various individuals and organizations are expected to play managerial roles in both hands-off (e.g., institutional design, goal, and framework steering) and hands-on (e.g., process management to encourage direct participation) ways. For example, policy-makers engaged in high-level decision-making are expected to take on a hands-off role, whereas middle-level public managers are expected to engage in hands-on activities (Torfing et al. 2012, 135).

As another important perspective in terms of interactions between government and nongovernmental actors, we should pay attention not only to effectiveness and efficiency but also to democratic aspects of interactive governance, such as accountability, transparency, and legitimacy. In terms of legitimacy, Torfing et al. (2012, 61) argue that interactive governance has "three important sources" of legitimacy: "input legitimacy," "throughput legitimacy," and "output legitimacy." Input legitimacy emphasizes "the composition of the participants," which is relevant to the basic rule of representative democracy in western developed countries. Throughput legitimacy derives through "showing the internal policymaking processes" that are expected to contain "commonly accepted normative ideals about fairness, responsiveness, and transparency." Output legitimacy derives "from providing desirable, useful, and promising policy solutions or from earning a reputation as a successful mechanism of governance or trouble-shooting." Torfing et al.'s discussion implies that we should pay attention to multifaceted sources of legitimacy in interactive governance relating to the existing formal governance system. Even by experiences and practices in democratic developed countries in Europe, "to what extent are civic initiatives (considered to be) legitimate" is still questionable; this issue is to be examined as one important future research question in interactive governance (Edelenbos and van Meerkerk 2016).

6 Interactive Approaches to Case Studies in Asia: Outlook of the Book Chapter

This book explores interactive approaches to case studies on water governance in Asia while referring to preceding water governance frameworks as well as the case in the Netherlands where we can find interesting literatures on interactive governance in the field of water management. We focus on cases of multi-stakeholder water governance issues in each country and analyze its interactive process among stakeholders as well as the process between formal (existing) and informal (emerging) institutions. Also, we look into broad political, economic, social, and cultural contexts behind each case. Focusing on interactions mentioned above under complex, contingent social-ecological systems with unique local contexts makes us rethink to what extent the concept of IWRM and its descendent frameworks on water governance are effective and what kinds of forms in terms of interactive perspectives as introduced in the previous section can be found in cases in Asian countries while referring to preceding cases in the Netherlands in terms of interactive form and process on water governance.

Before we step into the contents of case studies, we have to clear the wordings related to interactive "X"s. "Interactive governance" means an interactive form of governance generally and also means the core framework developed mostly in Western democratic countries in some contexts. Comparing to this seminal wording, "interactive approaches" and "interactive perspectives" mean methodology and sub-concepts referring to the theory of interactive governance which are applied or mentioned in case studies in different contexts. The "interactive process" means a kind of interaction seen in the process of water governance generally, and "interactive participation" is used to describe one interactive form in water governance which implicates a newly emerging public participation.

As a secondary introduction chapter including case studies in the Netherlands, one of Western democratic countries where we can find interesting cases on water governance, Chap. 2 reviews preceding cases of citizen engagement in water governance in the country, focusing on the interaction and relationship between citizen initiatives and governmental agencies in two cases from the Room for the River program, in order to provide a reference for Asian cases depicted in the following chapters. This program was initiated by the Dutch national government in 2006 under the Room for the River Directive, which was established in response to the massive flood disasters in the 1990s to integrate water management into spatial development. In the Netherlands, the traditional sectoral engineering approach has been used for water management for many years. However, the transition from this traditional method to an interactive method for integrated water management coproduced by citizens and government has been accelerated. Through comparative studies of two cases of flood risk management projects, this chapter shows contextual differences in the strategy, resources, and goals of initiators and governmental responses to stakeholder initiatives. The findings and discussions in this chapter contribute to the development of the theory of water governance and provide a perspective to analyze Asian cases where the role of government is still dominant in water governance.

Chapter 3 discusses the first Japanese case of dam removal policy. The Arase Dam in Kumamoto Prefecture was built for hydropower generation as a symbol of economic development in 1955. After its construction, local residents came to realize the damage caused by the dam, such as flood damage and eutrophication of the reservoir. In parallel with the anti-dam construction movement, at another site in the same river basin, the local village and fishermen began to ask the prefectural government to remove the Arase Dam when the permit for hydropower generation expired in 2003. Until the governor of the prefecture decided to remove the dam, there had been a series of complex interactions among different levels of government, local residents, fishery cooperatives, and downstream farmers, with political dynamics from changes of top leaders in elections. This chapter focuses on the contextual factors in interactive governance and stresses the importance of resistance strategy, which has been discussed in case studies of local commons in Japan, rather than collaborative governance discussed in many water governance studies.

Chapters 4 and 5 discuss cases in China, and both illustrate the process of participatory governance, including roundtable meetings, which are organized by multistakeholders to promote dialogue to address water and other environmental issues under the authoritarian regime led by the communist party.

Chapter 4 focuses on several types of participatory mechanism currently observed in watershed management in Zhejiang Province. The Watershed Roundtable Mechanism is a new mechanism with aims similar to that used for Taihu Lake Basin initiated by an expert team (discussed in Chap. 5); however, there are now diverse types of roundtable meetings in the province. In particular, after the provincial water policy called the Five Water Collaborative Governance was issued, new forms of roundtable meetings have been launched. The Our Water Roundtable, which was organized by an environmental nongovernmental organization (NGO) in Hangzhou City, and the Tiesha River Roundtable, organized by an environmental voluntary group founded by the Environmental Protection Bureau and the Communist Youth League of Hangzhou City, focused on the river issues. This chapter discusses how to lead successful interactive participation, which means one interactive form emerged in this region, beyond the consultative authoritarianism in China, and reviews the processes and characteristics of these roundtables.

Chapter 5 focuses on the pilot project initiated jointly by Japanese and Chinese research institutes in some communities in an industrial development zone in Yixing City, Jiangsu Province, which is located in the lakefront of Taihu Lake Basin. Here, the large coastal city, Wuxi City, experienced a drinking water crisis due to a huge bloom of blue-green algae in 2007. Subsequently, state, provincial, and local governments took more intensive measures to control water pollution in the basin. However, such top-down governance requires a bottom-up mechanism to be sustainable and effective in the long term. The joint research team conducted eight meetings with the cooperation of the local community leader and grassroots government in their pilot project and identified achievements and difficulties in initiating a bottom-up, interactive mechanism without an official institution. This chapter focuses on the issue of legitimacy in its discussion and introduces the concept of institutional legitimacy to address problems in promoting interactive governance under the authoritarian regime in China.

Chapter 6 focuses on two cases of people's organizations in water governance in northern Thailand. This chapter aims to reveal the role of civil society as a stakeholder in water management, including water allocation and flood prevention of the Mekong's tributaries in Chiang Rai. We clarify the interaction and coordination between local NGOs and government authorities, focusing on participatory opportunities and negotiation capacity. We examine the two case studies of the People Council of Ing River and water allocation and the Association of Chiang Saen Livable City and Kok River Basin Ecology Group and flood prevention by telemetry and early warning systems. This chapter argues the limitations of local NGOs' participation and their negotiation with government authorities in interactive, cooperative way of water governance.

Chapter 7 deals with the watershed and water resource management of Laguna Lake in the Philippines. Laguna Lake is the largest freshwater lake in the Philippines and is located next to the capital, Metropolitan Manila. Due to urbanization and industrialization, the lake's water quality is deteriorating, and water resource management is now urgently required. This chapter reviews the Philippines' water governance and its history and framework and then examines the activities and roles of stakeholders for Laguna Lake. We focus on the Laguna Lake Development Authority and the role of the Yaman ng Lawa initiative in local community activities for managing and using the lake and its watersheds sustainably. This chapter discusses how community mechanisms can produce social capital by interactive governance in the lake basin.

Chapter 8 examines the process and outcomes of action research on irrigation management in Sulawesi, Indonesia. The objectives of the research were to provide an opportunity for old and new stakeholders to meet, discuss, and make decisions together and to provide experience of using interactive processes in implementing agreed-upon solutions. By jointly experiencing challenges and successes based on mutual decision-making with moderate support from the researchers, a space for participation and creating contextualized roles and responsibilities was formed based on the sociocultural situation. Moreover, this occurred regardless of whether the participant was on the "governing" or "governed" side. This chapter addresses the dynamic integration of local customs and government-led systems facing mutual discontinuity, which is a concern from the cultural perspective of interactive governance.

Chapter 9 focuses on the physical, institutional, and community characteristics and their interrelations which drive the interactive governance of Ana Sagar toward its sustainability. Ana Sagar is located in the heart of Ajmer City in Rajasthan which is a pilgrim city. The lake has been historically a rainwater harvesting reservoir; however, in today not only rainwater but also wastewater is running into the lake. In the IAD framework adapted in this chapter, it is unique to consider the institutional characteristics such as the policies, the governance activities, and the stakeholders and also the structure that encompasses the contextual characteristics such as the biophysical conditions of the lake and the community attributes while examining interactions and outcomes in lake governance. In the recent lake restoration process, the decided full tank water level of the lake has led to submergence of several developments at the lakeshore. This has led to dissatisfaction among the local people since several of the submerged lands belong to them. Ana Sagar is an atypical lake in India whose lake area is divided into several land parcels and ownerships including both private and government. The water pollution continues in the lake since the inlet drainages carrying rainwater and wastewater from the surrounding hills and urban development are awaiting treatments before flowing into the lake. Facing the dilemma of development and conservation of the lake by the physical, institutional, and community factors, the chapter discusses the complex problems and processes that influence the lake governance toward sustainable development.

In the last concluding section, we synthesize the findings and discussions from all the chapters in this book and identify the remaining tasks for interactive approaches to water governance in Asian cases.

7 Conclusion

In this book, we introduce interactive perspectives, which have been discussed mainly in the context of Western European countries, to case studies on water governance in Asia. We examine how these perspectives reveal complex and dynamic interactions in water governance in Asia and how interactions between policies and practices as well as interactions between formal institutes and emerging informal institutes occurred. This section will synthesize findings and discussions from seven case studies in Asia referring to those in the Netherlands as preceding cases of interactive governance for integrated water management.

7.1 A Summary of Case Studies

Here we can summarize our findings of case studies on water governance in Asia from interactive perspectives while addressing remained tasks to be investigated further as below.

First, in most countries where we conduct case studies, water resource and environmental management is still occupied with governmental sectors; however, they cannot solve problems effectively alone under hierarchical and fragmented administrative system (Chaps. 3, 4, 5, 6, and 8).

Second, under such a situation, some interactive forms of governance have been emerging, and those forms are not initiated by citizens to be seen in the case in the Netherlands (Chap. 2), but resisted or involved by people; this is because the government which is not tolerant to citizen-initiative form of governance (Chaps. 3, 4, and 5) has still decisive power and also because civic organizations are too fragmented to negotiate with the government (Chap. 6).

Third, it should be noted that hybrid forms of interactive governance including governmental and nongovernmental actors have been emerging in some cases, and under such forms, the expected role of the government as well as nongovernmental actors could be changed in more cooperative way to solve the problems (Chaps. 4, 7, and 8). In some cases, researchers outside the locality could play an important role to facilitate such a hybrid form of interactive governance especially in its initial stage (Chaps. 7 and 8). The role of such a transdisciplinary approach should be investigated in other cases in Asia where there are still large gaps of governance capacity between countries.

Fourth, we can find the important role of a street-level bureaucrat who can conduct hands-on intervention to interactive governance, while he/she can be involved in the interactive process. Such a dual role of the street-level bureaucrat is mentioned only in Chap. 8 explicitly; however, we can find it in other chapters implicitly (Chaps. 4, 5, 6, 7, and 9). This should be investigated further in other cases in Asia.

Fifth, it should be mentioned that contextual factors are significant when we compare cases not only between countries but also within the same counties (Chaps. 4 and 5). This is significant beyond different regions as well as different institutional situations.

Lastly, complexities of social and ecological systems (SESs) in river and lake basins could pose us to rethink the way of governing issues appearing in boundaries between nature and human system (Chaps. 6 and 9). Especially in the Indian case (Chap. 9), we can find diversified and intimate relations within SESs. Under such complex and diversified characteristics, it should be investigated further how we can coproduce an adaptive way of governing water issues as well as sustainable development from interactive perspectives in each case.

Following these summaries, we discuss some topics of interactive perspectives on water governance in more details as below.

7.2 A Transition in Water Governance?

In the Netherlands, as Chap. 2 shows, citizens take initiative in all kinds of sectors, including water and flood risk management. In particular, in water and flood risk management, there has been a cautious transition from top-down and expert-led decision-making to network coordination and stakeholder approaches, largely due to the empowerment of interest groups and citizens by professionalization. The Room for the River program is one such case, which requires involvement of local and regional stakeholders to formulate and decide plans and designs under the national governmental framework. Under this program, flood control facilities, such as dykes and breakwaters, have been relocated or downsized through spatial planning.

A similar transition has occurred in Japan. As Chaps. 1 and 3 mention, Japan has reformed its water-related policy framework since the 1990s. In 1997, the River Law was amended to address the function of environmental conservation in the rivers in addition to traditional functions including flood control and water use. The revised law requires river administrators to listen and incorporate the voices of concerned parties in creating basic plans, and river basin committees have also been set up to incorporate these voices in plans. However, as Chap. 1 points out, several river development projects, such as dam constructions, have been advanced by government agencies despite strong opposition and protests by inhabitants who suffer the negative effects of the projects. The Arase Dam removal analyzed in Chap. 3 is a pioneering case ending the water engineering-centered era, which has dominated in Japan for several decades, similar to the Netherlands.

Thus, the transition from an engineer-centered paradigm to a multi-stakeholder paradigm has been observed in Japan. In other Asian countries, although we observe an emerging interactive form of governance, it is unclear whether this transition for water governance has already occurred or will occur soon, due to the limitations of our case studies. This is a question that we should pursue further. Here, we can only say that we observe a variety of emerging forms of interactive governance in the field of water management in many counties in Asia, whether or not a transition in water governance has occurred.

7.3 Contextualization

As discussed in Chap. 1, integrated water resource management (IWRM) depends on the local context, which determines how to select and organize tools for solving water management problems locally even though those tools are effective in other cases. The instruments in IWRM and its descendent frameworks on water governance are not panaceas for water problems in any part of the world. Even within Western European countries and the USA, which have many similarities in their democratic traditions, the water management transition from top-down management to more egalitarian forms of multi-actor network governance differs between countries. The transition depends on historical, geographical, and institutional contexts encompassing diversity in physical, political, cultural, historical, and policy factors, as suggested in Chap. 2.

The case study on the Arase Dam removal policy in Chap. 3 demonstrates that contextual factors are important in determining the mode of governance in developing multi-stakeholder engagement in public policy. In this case, two contextual factors affect the mode of governance. First, there was a clear power imbalance between those who advocated the dam's removal and those who advocated the status quo. Second, there were different policy beliefs among stakeholders, resulting in their confrontational relationships. Thus, in this case, local resistance against the prefectural and national government emerged instead of a collaborative relationship, which has often been discussed in many water governance studies, and this resistance strategy pushed forward the dam removal policy.

The issue of legitimacy in interactive governance appears in a different sociopolitical context, as discussed in Chap. 5. A seminal book about interactive governance theory (Torfing et al. 2012) argues that interactive governance has three sources of legitimacy: input legitimacy, throughput legitimacy, and output legitimacy. Based on a case study on the pilot project on community roundtable meetings in Taihu Lake Basin, China, Chap. 5 points out that institutional legitimacy is critical to organizing interactive dialogue under the fragile formality of meetings under the authoritarian regime in China. This is another type of legitimacy which should be studied further in other authoritarian countries.

It is interesting to compare the sociopolitical context of the two cases in China in Chaps. 4 and 5. In the case study of Zhejiang Province in Chap. 4, there are diverse participatory governance mechanisms for water management, and some types of roundtable meetings have been held to promote a dialogue among stakeholders. In this province, local people do not face institutional legitimacy explicitly because this form of interactive governance was introduced and is supported by official media, which is a propaganda institution controlled by the communist party. Thus, we can understand the importance of sociopolitical contexts nationally as well as locally in affecting the form of interactive governance in China.

7.4 (Un)changing Roles of Government

The role of government and nongovernmental stakeholders, including citizens in terms of Western European concept, is a key factor in understanding the process and effectiveness of interactive governance. According to a critical review of interactive governance in Europe, Edelenbos and van Meerkerk (2016) categorized interactive governance as government-induced interactive governance and citizen-induced interactive governance. As reviewed in Chap. 2, the Netherlands has three ways of generating citizen engagement in public policy: formal public participation, interactive policy-making, and citizens' initiatives. Through this development of interactive governance, the roles of stakeholders, including those of the government and citizens, are expected to change. In the Netherlands, where interactive governance has developed from government-induced to citizen-induced, all types of generation of citizen engagement are now observed, and there are combined centralized and decentralized government roles.

In Japan, where there is a transition in water management similar to the Netherlands, there is a change in the attitude of government for some issues, as in the Arase Dam removal described in Chap. 3. However, there could be no change in the role of government, which still has the exclusive power to decide whether to build or remove water facilities. This is a contextual factor that induces the resistance strategy rather than the cooperative strategy.

This mixed change in the attitude and role of the government is also seen in the pilot project on community roundtable meetings in Taihu Lake Basin, China, as discussed in Chap. 5. Although there are dialogues between residents, enterprises, and government, local government has changed its attitude to solving a local problem highlighted by residents; this does not mean that government has changed its decisive role in local governance. The case studies in Chap. 2 show a similar situation in the Netherlands; however, the decisive role of the government in Japan and China in their case studies seems more robust than that in the case of the Netherlands.

Another important role of government in interactive governance is metagovernance, which is steering and promoting the interactive process by hands-off or hands-on methods, as discussed in Chap. 1. In the case in Chap. 5, we can see the hands-on intervention by a provincial official to legitimize the community roundtable meetings under the fragile formal setting. This tells us implicitly such a meeting could be steered from the ruling party and the government in a hands-off way under the social stability first policy of the authoritarian regime in China. Contrary to this observation, we can also see the changing role of the government in Indonesia as discussed later. It should also be noted that hands-on interventions by bureaucrat in the lake restoration program are causing its fragmented implementation in the case of India (Chap. 9). Thus, the form and effect of metagoverance are more complicated than expected.

7.5 *Limited Roles of Nongovernmental Stakeholders*

The case studies in Asia in this book demonstrate the mixed roles of nongovernmental stakeholders in water management, including resistance and cooperation. In the case in Japan (Chap. 3), there is resistance by the local fishery cooperative and inhabitants. In the case of India (Chap. 9), there is a conflict between residents and local government over compensation for the lake submergence. In the cases in China (Chaps. 4 and 5) and in Laguna Lake in the Philippines (Chap. 7), they discuss the cooperative role of residents in water and environmental governance.

In Chap. 6, the mixed roles of local people are discussed for the people's council in Northern Thailand. Thailand has a centralized administrative system that gives exclusive power to the government and a fragmented structure with no effective coordinative function in the existing river basin committee. In the absence of an effective IWRM mechanism, people's councils have been set up in the Ing River based on several community networks, which are grassroots organizations that fight for their survival and subsistence. These community networks have resisted and negotiated with the government against top-down development plans with no informed consent from local people.

These people's councils are now seeking to cooperate with the government, and the government also expects their cooperation; however, there are perception gaps about the causes of and solutions to environmental degradation in the river basin. The government thinks that the problem can be solved by engineering and technical methods, whereas the people's councils think it could be solved by alternative development for sustainable livelihoods with environmental conservation. Another case in the Kok River Basin (Chap. 6) demonstrates the difficulty in people's organizations cooperating with the government and the fragmentation within people's organizations in the same area for dealing with floods caused by the dam and navigation management in the upper stream of the Mekong River in China.

Thus, in Northern Thailand, community self-organizations have emerged; however, their role is still limited by the centralized, fragmented administrative system and fragmentation within people's organizations. This case implicates the right-based approach as mentioned in the conclusion of this chapter will be needed to balance interactions between people and the government in river basin governance.

7.6 *Hybrid Form: A Coproduction in Asia*

Concerning multi-stakeholder involvement in water governance in Asian cases, we can find a hybrid form of governance consisting of governmental and nongovernmental actors, whether it is led by government or not in some cases. In Chap. 4, the case study in Zhejiang Province shows several hybrid forms of interactive participatory mechanism as roundtable meetings organized by stakeholders, including governmental and nongovernmental sectors, and mixed sectors, such as local government, TV stations, and voluntary organizations led by the government. The Tiesha River Roundtable is an important case that continues in an independent and sustainable manner; however, the roundtable focuses only on exchanging experiences among participants, not on problem-solving related to policy-making with no institutions to incorporate the information into official plans.

Another type of hybrid form is found in the action research in Indonesia in Chap. 8. The action research has been conducted jointly by local and international researchers to improve irrigation management in South Sulawesi, which has been fragmented between officials, farmers, traditional water managers (called MJ), and government-led modern farmers' organizations (called P3A). Through several years of field research, seasonal gate control and water distribution have been improved by a documentation method and scientific data being collected and produced jointly by researchers and local stakeholders and shared among prefectural and local stakeholders. This action research has connected governmental policies with daily farming practices and modern water management with traditional methods beyond cognitive boundaries by internalizing organizational values in the local context. Furthermore, the roles of federation leaders and technicians have changed throughout the action research to learning local contexts for irrigation management. This is a good example of coproduction by local stakeholders and international researchers as a hybrid form composed of governmental and nongovernmental, central and local, modern and traditional, domestic and international personnel, and researches and practices.

There is another example in the case of Laguna Lake in the Philippines, discussed in Chap. 7. In this case, local and international researchers intervened in lake governance to ensure sustainable use of lake resources, and local stakeholders have been seeking a way to develop their experiences learned in joint activities. This is an emerging methodology called "transdisciplinary approach" to address problem-solving issues to cooperate not only beyond academic disciplines (often called as "interdisciplinary") but also with nonacademic practitioners (Brandt et al. 2013). Like those cases in Southeast Asian countries (Chaps. 7 and 8), transdisciplinary intervention in hybrid form of interactive governance can play a significant role to empower local people and other nongovernmental stakeholders who have limited roles to be involved in coproduction.

7.7 *Toward Sustainable Development Embedded in Social and Ecological Systems*

In considering interactions between social actors and between human communities and the natural environment, interactive perspectives in water governance should be integrated with ecological perspectives, which have been developed in the research on the commons and basin governance as its descendent framework introduced in Chap. 1. This thinking can lead to the social and ecological system perspectives applied to the case study of Ana Sagar, an urban lake in India (Chap. 9). In this case, fostering urban development and protecting the health of the lake ecosystem are a big challenge. Ana Sagar was used as a rainwater harvester but now is a wastewater reservoir owing to the urban development surrounding the lake and the activities of locals and visitors. Dynamic changes in the water level of the lake and multiple property rights to the land surrounding the lake make the lake governance complex and difficult. In the restoration plan, there is no effective and integrated management for this issue, and there is insufficient deliberation between local governments and local urban communities. Incorporating biophysical and social factors into social and ecological systems is useful for dealing with complex governance issues and is also important for achieving lake sustainability.

The sustainability of rivers and lakes is a common challenge in most parts of Asia because of the threat of urbanization, industrialization, lifestyle modernization, and other factors caused by natural environmental change and human activities like climate change. For example, Laguna Lake in the Philippines (Chap. 8) faces water pollution, fish kills, and other environmental changes that threaten its sustainability. The Ing and Kok Rivers in Northern Thailand have also been facing intended and unexpected water fluctuation due to mixed factors caused by climate change and human activities in the upper stream. In these cases, social and ecological system perspectives could be incorporated into interactive perspectives to understand the social and ecological complexities and their dynamics for sustainable development to improve people's subsistence and welfare while seeking wise water governance.

8 Appendix: Tools for IWRM

A. Enabling environment
A1. Policies
A1.01 Preparation of a national water resources policy
A1.02 Policies with relation to water resources
A1.03 Climate change adaptation policies
A2. Legal framework
A2.01 Elements of water law
A2.02 Implementation and enforcement
A2.03 Integrated legal frameworks for IWRM
A3. Investment and financing structures
A3.01 Investment frameworks
A3.02 Strategic financial planning
A3.03 Generating basic revenues for water
A3.04 Repayable sources of finance for water
B. Institutional roles
B1. Creating an organizational framework
B1.01 Reforming institutions for better governance
B1.02 Transboundary organizations for water resource management
B1.03 National apex bodies
B1.04 River basin organizations
B1.05 Regulatory bodies and enforcement agencies
B1.06 Service providers and IWRM
B1.07 Strengthening the public sector water utilities
B1.08 Role of the private sector
B1.09 Civil society institutions and community-based organizations
B1.10 Local authorities
B1.11 Building partnerships

(continued)

B2. Building institutional capacity
B2.01 Participatory capacity
B2.02 Capacity of water professionals
B2.03 Regulatory capacity
C. Management instruments
C1. Water resources assessment
C1.01 Water resources knowledge base
C1.02 Water resources assessment
C1.03 Modeling in IWRM
C1.04 Developing IWRM indicators
C1.05 Ecosystem assessment
C1.06 Water footprint and virtual water concept
C2. Plans for IWRM
C2.01 National IWRM plans
C2.02 Basin management plans
C2.03 Groundwater management plans
C2.04 Coastal zone management plans
C2.05 Water infrastructure implementation and IWRM
C3. Efficiency in water use
C3.01 Efficiency of use
C3.02 Recycling and reuse
C3.03 Efficiency of supply
C4. Social change instruments
C4.01 Education curricula on water management
C4.02 Communication with stakeholders
C4.03 Raising public awareness
C5. Conflict resolution
C5.01 Conflict management
C5.02 Shared vision planning
C5.03 Consensus building
C6. Regulatory instruments
C6.01 Water rights and allocation
C6.02 Water quality
C6.03 Water services
C6.04 Land use
C6.05 Protecting freshwater ecosystem resources
C7. Economic instruments
C7.01 Pricing of water and water services
C7.02 Pollution charges
C7.03 Water markets and tradable permits
C7.04 Subsidies
C7.05 Payment for environmental services

(continued)

C8. Information exchange	
	C8.01 Information management systems
	C8.02 Sharing data for IWRM
C9. Assessment instruments	
	C9.01 Risk assessment and management
	C9.02 Environmental assessment
	C9.03 Social assessment
	C9.04 Economic assessment
	C9.05 Vulnerability assessment

Source: Global Water Partnership IWRM Toolbox, www.gwp.org/en/ToolBox/TOOLS/Management-Instruments/Assessment-Instruments/, accessed 29 January 2016

References

Berkes, Fikret, Johan Colding, and Carl Folke, eds. 2003. *Navigating Social-Ecological Systems: Building Resilience for Complexity and Change*. Cambridge: Cambridge University Press.

Brandt, Patric, Anna Ernst, Fabienne Gralla, Christopher Luederitz, Daniel J. Lang, Jens Newig, Florian Reinert, David J. Abson, and Henrik von Wehrden. 2013. A Review of Transdisciplinary Research in Sustainability Science. *Ecological Economics* 92: 1–15. https://doi.org/10.1016/j.ecolecon.2013.04.008.

Dietz, Thomas, Elinor Ostrom, and Paul C. Stern. 2003. The Struggle to Govern the Commons. *Science* 302 (5652): 1907–1912.

Edelenbos, Jurian. 2005. Institutional Implications of Interactive Governance: Insights from Dutch Practice. *Governance: An International Journal of Policy, Administration, and Institutions* 18 (1): 111–134.

Edelenbos, Jurian, and Geert Teisman. 2013. Water Governance Capacity: The Art of Dealing with a Multiplicity of Levels, Sectors and Domains. *International Journal of Water Governance* 1 (1/2): 89–108.

Edelenbos, Jurian, and Ingmar van Meerkerk. 2016. *Critical Reflections on Interactive Governance: Self-organization and Participation in Public Governance*. Cheltenham/Norhampton: Edward Elgar.

Edelenbos, Jurian, Nienke van Schie, and Lasse Gerrits. 2010. Organizing Interfaces Between Government Institutions and Interactive Governance. *Policy Sciences* 43: 73–94.

Edelenbos, Jurian, Nanny Bressers, and Peter Scholten. 2013. *Water Governance as Connective Capacity*. London: Ashgate.

Fujita, Kaori. 2005. Evaluating Cost Sharing for Sustainable River Basin Management: Case Studies in Netherlands and Japan. In *Promoting Sustainable River Basin Governance: Crafting Japan-U.S. Water Partnerships in China*, IDE Spot Survey 28, ed. Jennifer L. Turner and Kenji Otsuka, 103–122.l. Chiba: IDE-JETRO.

———. 2015. Nousanson no ijikanousei to genkai syuraku mondai heno taiou: Kochi-ken Niyodogawa-cyo no jirei kara [Sustainability of Mountain Rural Villages and Responses to Marginal Community Problems: Case Study in Niyodogawa Town, Kochi Prefecture in Japan]. In *Ajia no Seitaikiki to Jizokukanousei: "Field" kara no "Sustainability" ron* [Ecological Crisis and Sustainability in Asia: A Synthesis of Field Studies], ed. Kenji Otsuka, 149–190. Chiba: Institute of Developing Economies. (Japanese).

GWP-TAC (Global Water Partnership Technical Advisory Committee). 2000. *Integrated Water Resources Management*. Stockholm: GWP.

GWP-TEC (Global Water Partnership Technical Committee), Peter Rogers, and Alan W. Hall. 2003. *Effective Water Governance*. Stockholm: GWP.

ILEC. 2005. *Managing Lakes and Their Basins for Sustainable Use: A Report for Lake Basin Managers and Stakeholders*. Kusatsu-shi: International Lake Environment Committee Foundation (ILEC).

Lubell, Mark, and Jurian Edelenbos. 2013. Integrated Water Resources Management: A Comparative Laboratory for Water Governance. *International Journal of Water Governance* 1 (03–04): 177–196.

MEA (Millennium Ecosystem Assessment). 2005. *Ecosystems and Human Well-being: Synthesis*. Washington, DC: Island Press.

Mitsumata, Gaku. 2013. Complementary Environmental Resource Polices in the Public, Commons and Private Spheres: An Analysis of External Impacts on the Commons. In *Local Commons and Democratic Environmental Governance*, ed. Takeshi Murota and Ken Takeshita, 40–65. New York: United Nations University Press.

Murota, Takeshi, and Ken Takeshita eds. 2013. *Local Commons and Democratic Environmental Governance*. New York: United Nations University Press.

OECD. 2015. *OECD Principles on Water Governance*. Paris: OECD.

Ostrom, Elinor. 1990. *Governing the Commons: The Evolution of Institutions for Collective Action*. Cambridge: Cambridge University Press.

———, ed. 2015. *Ajia no Seitaikiki to Jizokukanousei: "Field" kara no "Sustainability" ron* [Ecological Crisis and Sustainability in Asia: A Synthesis of Field Studies]. Chiba: Institute of Developing Economies (Japanese).

Pangare, Vasudha, Ganesh Pangare, Viraj Shah, B.R. Neupane, and P. Somaekhar Rao. 2006. *Global Perspectives on Integrated Water Resources Management: A Resource Kit*. New Delhi: Academic Foundation.

RCSE-SU and ILEC. 2014. *Development of ILBM Platform Process; Evolving Guidelines Through Participatory Improvement*, 2nd ed. Research Center for Sustainability and Environment, Shiga University (RCSE-SU) and International Lake Environment Committee Foundation (ILEC).

Teisman, Geert, Arwin van Buuren, Jurian Edelenbos, and Jeroen Warner. 2013. Water Governance: Facing the Limits of Managerialism, Determinism, Water-Centricity, and Technocratic Problem-Solving. *International Journal of Water Governance* 1 (1): 1–11.

Torfing, Jacob, B. Guy Peters, Jon Pierre, and Eva Sørensen. 2012. *Interactive Governance: Advancing the Paradigm*. Oxford: Oxford University Press.

Turner, Jennifer L., and Kenji Otsuka, eds. 2005. *Promoting Sustainable River Basin Governance: Crafting Japan-U.S. Water Partnerships in China*, IDE Spot Survey, 28. Chiba: IDE-JETRO.

UNDP. 2006. *Human Development Report 2006–Beyond Scarcity: Power, Poverty and the Global Water Crisis*. New York: UNDP.

WWAP (United Nations World Water Assessment Programme). 2003. *Nations World Water Development Report: Water for People, Water for Life*. Paris: UNESCO.

———. 2006. *Nations World Water Development Report 2: Water, a Shared Responsibility*. Paris: UNESCO.

———. 2009. *Nations World Water Development Report 3: Water in a Changing World*. Paris: UNESCO.

———. 2012. *Nations World Water Development Report 2012: Managing Water under Uncertainty and Risk*. Paris: UNESCO.

———. 2015. *The United Nations World Water Development Report 2015: Water for a Sustainable World*. Paris: UNESCO.

Kenji Otsuka is Senior Research Fellow in the Environment and Natural Resource Studies Group at the Interdisciplinary Studies Center, Institute of Developing Economies, Japan External Trade Organization (IDE-JETRO), Chiba, Japan. His research interests are in the fields of environmental issues, policies, and social change in China as well as interactive governance on water, environment, and health and disaster risk reduction in Asia. He has been organizing joint research projects with Chinese counterparts, including both governmental research institutes and NGOs since 1993.

Chapter 2
Citizen Initiatives in Water Governance in the Netherlands: Reflection and Implication to Asian Cases

Jurian Edelenbos

Abstract This chapter reviews preceding cases of citizen engagement in water governance in the country, focusing on the interaction and relationship between citizen initiatives and governmental agencies in two cases from the Room for the River program, in order to provide a reference for Asian cases depicted in the following chapters. In the Netherlands, the traditional sectoral engineering approach has been used for water management for many years. However, the transition from this traditional method to an interactive method for integrated water management co-produced by citizens and government has been accelerated. Through comparative studies of two cases of flood risk management projects, this chapter shows contextual differences in the strategy, resources, and goals of initiators and governmental responses to stakeholder initiatives. The findings and discussions in this chapter contribute to the development of the theory of water governance and provide a perspective to analyze Asian cases where the role of government is still dominant in water governance.

Keywords Citizen initiative · Interactive governance · Water management · Spatial planning · Government support

1 Introduction

In this chapter, interactive governance in water management in the Netherlands (Western Europe) is discussed. Specifically, the emergence of citizen initiatives as a form of interactive governance is described and analyzed. This implies that the emphasis will be on the new forms of citizen engagement, that is, the way that citizens take their own responsibility and develop activities to prepare and implement a public good or service (Edelenbos and Van Meerkerk 2016). In the Netherlands, we see that citizens take initiative in all kinds of sectors, such as water and flood risk management.

J. Edelenbos (✉)
Department of Public Administration & Sociology, Erasmus School of Social and Behavioral Sciences, Erasmus University Rotterdam, Rotterdam, The Netherlands
e-mail: edelenbos@essb.eur.nl

K. Otsuka (ed.), *Interactive Approaches to Water Governance in Asia*,
https://doi.org/10.1007/978-981-13-2399-7_2

This new form of citizen engagement, as part of interactive governance, has consequences for the role of governments in public policy-making, specifically in the field of water management in which the Netherlands has a long tradition (Lintsen 2002). In this chapter we focus on the interrelationship between citizen initiatives on the one hand and the (changing) role of governmental agencies on the other hand. We investigate to what extent the interaction and relationship between citizen initiatives and governmental institutions predict the level of impact that citizen initiatives have in the field of water management.

In order to meet this goal, we have structured the chapter in the following ways. First, we pay attention to institutional backgrounds of the Netherlands: the governance structure and the emergence of interactive governance, specifically citizen initiatives. We also pay attention to the tradition of water management in the Netherlands, developing from an engineer approach toward integral and adaptive water management approach leading to the start of the Room for the River program. Second, we develop a theoretical framework, elaborating key concepts of citizen initiative, governmental response to citizen initiatives, and impact, in order to study in depth two case studies from the Dutch Room for the River program. We shortly discuss methodological issues. Third, we analyze and discuss the case study results. We finalize the chapter with key conclusions and reflection and implication to Asian cases.

2 Background

2.1 Netherlands Governance Structure and Emergence of Interactive Governance

2.1.1 General Institutional Information

The politics of the Netherlands take place within the framework of a parliamentary representative democracy, a constitutional monarchy, and a decentralized unitary state. Citizens choose their (political) representatives once every 4 years, and the governmental tasks and accountabilities are distributed among three layers of the government, the national, regional, and municipal level. Furthermore, the Netherlands is described as a consociational state (Lijphart 1969). Dutch politics and governance are characterized by a common striving for broad consensus on important issues, within both the political community and society as a whole (Lijphart and Aitkin 1994). The Dutch Parliament or States General consists of a Lower House or Second Chamber and an Upper House or First Chamber, also referred to as the Senate.[1] Regional government in the Netherlands is formed by 12 provinces. Provinces are responsible for spatial planning, health policy, and recreation, within the bounds

[1] Both houses of Parliament discuss proposed legislation and review of the actions of the cabinet. The Second Chamber also has the right to propose or amend legislation.

prescribed by the national government. Furthermore, they oversee the policy and finances of municipalities and water boards.

Local government in the Netherlands is formed by 418 municipalities. Municipalities are responsible for education, spatial planning, and social security, within the bounds prescribed by the national and provincial government. They are governed by the College of Mayor and Aldermen. The Mayor is appointed by the national Cabinet and responsible to the Minister of the Interior and Kingdom Relations. The Aldermen are appointed by and responsible to the Municipal Council, which is elected by direct suffrage (Edelenbos et al. 2016).

Furthermore, there are water boards which are responsible for the country's polders, dikes, and other waterworks. These bodies are elected in nonpartisan elections and have the power to tax their residents (Van Buuren et al. 2012b).

2.1.2 Citizen Engagement and the Emergence of Citizen Initiatives

In most democracies a form of representative democracy is chosen, because a direct form of political decision-making on all policy issues by all citizens would be unworkable. That doesn't change the fact that almost all existing democracies are a combination of both direct and indirect democracy (Mayer et al. 2005), for example, as result of the introduction of citizen initiatives and referenda in a parliamentary system.

In recent strands of governance theory, there is special focus on *interactive* governance (Kooiman 1993; Edelenbos 2005; Torfing et al. 2012; Edelenbos and Van Meerkerk 2016), focusing on the interactions and initiatives of a plurality of public, societal, and private actors in dealing with complex societal issues, like water management. Although different constellations of public, private, and societal actors can be the locus of interactive governance, in this chapter we specifically aim at interactive forms of governance between governments and citizens. In this respect, interactive governance deals with civic engagement, stakeholder participation, self-organization, and civic/citizen initiatives (see also Edelenbos and Van Meerkerk, 2016).

Interactive forms of governance between citizens and government can be provoked by both citizens and government (Edelenbos and Van Meerkerk, 2016). Government-induced interactive governance generally refers to "citizen participation" and has become a significant policy strategy in many Western countries. It has been put, for example, at the heart of spatial planning, social care, regeneration, housing, and education policies (e.g., Irvin and Stansbury 2004). It is a form of interactive governance which is strongly organized by governments. This means that governments often decide when, who, and how people get involved. At certain moments in decision-making, governments initiate participation procedures in which citizens get a chance to respond to or provide input for decision-making highly structured by rules set by the government. Outcome from participation procedures is highly uncertain as the government can decide to a large extent what (not) to do with it. This often creates dissatisfaction among participants (Edelenbos 2005) and leads to reluctance to future engagement in "traditional institutions" or "participation procedures" of representative democracy (Sørensen and Torfing 2007).

Table 2.1 Three generations of citizen engagement in the Netherlands

	Definition	Period
1st generation	Formal public participation: statutory right for citizens to formally react (to speak) on intended decisions	Early 1970s until now
2nd generation	Interactive policy-making: citizens have the opportunity to use informal channels and in an early stage co-produce the policy	Early 1990s until now
3rd generation	Citizens' initiative: citizens determine the subjects (issues, ideas, plans) which will be decided by policy makers	2002 until now

Source: author

On the other hand, interactive forms of governance can be provoked bottom-up: by citizens themselves, i.e., citizen initiatives. Citizen initiatives can be initiated by residents, social entrepreneurs, artists, etc., as long as the initiative pursues a community purpose and not a direct business purpose. Several authors argue that this specific form of interactive governance is on the rise in many liberal democracies (e.g., Bang 2004; Dalton 2008; Marien et al. 2010). They reflect new forms of civic engagement, which can be labelled as self-organization (Boonstra 2015; Van Meerkerk et al. 2013). Although the phenomenon of bottom-up civic initiatives or citizen self-organization is not new, historical research shows that the current rise can be considered as a new "wave." For example, De Moor (2013) in the case of the Netherlands and Healey (2014) in the case of England refer to a previous wave at the end of the nineteenth century in which civil society initiatives emerge to address all kinds of needs arising from a rapidly expanding urban working-class population.

The citizen initiative can be seen as an addition from the participatory democracy to the existing representative democracy. The citizen initiative can be seen as a form of citizen engagement that fits into the third generation of citizen participation in the Netherlands in the twentieth and twenty-first centuries. Three generations of citizens' involvement can be distinguished (Edelenbos and Van Meerkerk 2016). Although we can speak of three generations, nowadays all forms of citizen engagement are present in the Netherlands (Table 2.1).

The first generation of citizen engagement, formal public participation procedures, was a response to the "social-political decompartmentalization" (Lijphart 1969) which was introduced in the 1960s.[2] In the first generation, citizens only had a reactive role; citizens could formally respond to (draft) decisions of governments. In the second generation, citizens got the opportunity to engage in the early stages of decision-making to give input that governments could use for reaching decisions. This form of participation was however completely created and controlled by the government. In this sense existing institutions of representative democracy do not have to adapt their processes and ways of working in a significant manner. In the

[2]This process means that the social, ideological, and religious pillars on which the Dutch society was built started to crumble. This process was accompanied with a process of democratization and emancipation of citizens. Citizens more and more resisted to hierarchical decision-making by political elites (Lijphart, 1969) and claimed more say in decision-making and personal freedom.

third generation, citizens can take initiative and put subjects on the political agenda. They can use the formal route that has been created by governments, but we also see that citizens take the informal route and present their ideas and plans informally to civil servants, administrators, and local politicians. This last form of participation could lead to governance arrangements between governmental organizations and self-organizing parts of the civil society. This asks for significant processes of adaptation within the existing institutions of representative democracy, because processes of decision-making and policy-making are altered.

Citizen initiatives can be approached as a process of self-organization where (organized) citizens and social interest groups spontaneously come to a common action (Van Meerkerk et al. 2013). Citizen participation is often initiated and regulated by the government (Edelenbos 2005); citizen initiative, however, is a bottom-up development started by citizens themselves (Edelenbos and Van Meerkerk 2016). This can be defined as a process of self-organization or self-regulation where (organized) actors, like citizens and social interest groups, spontaneously come to a common action (c.f. Rhodes 1997). Citizen initiatives often arise from dissatisfaction with the actions of governments and function as a response to proposed government policy. Citizens and social groups often see that resistance is useless and then switch to a more proactive way of resistance by developing plans on their own initiative. Processes evolve out of events, actions, and interactions and build an institutional structure (Teisman et al. 2009). Through interaction and bonding among citizens and public officials, information exchange, learning, and mutual experience develop, which may promote new patterns of relationships. Processes of self-organization in turn might lead to new relationships between governmental institutions and civil society.

2.2 The Netherlands: Water Management History and Tradition

For a long time, the Dutch governance system on water management was focused on ensuring a high level of safety by keeping water at a distance. For centuries, the Dutch invested a lot of money in safety measures against flooding by enforcing dikes, enlarging dams, and so on (Lintsen 2002; Rijke et al. 2012).

After the flood of 1953, this philosophy was further strengthened. Parliament agreed on the Delta Law (1 December 2011), where the Delta Works were proposed. As a result, nearly all sea arms were closed off. Dutch water management was characterized by a technocratic and state-oriented system of agencies like the water boards, provinces, and Directorate-General for Public Works and Water Management (Rijkswaterstaat, RWS), which is a part of the Ministry of Transport, in close cooperation with their own research institutes. A solid legal framework accompanies this paradigm and made powerful action in the water sector possible in many cases. Lintsen (2002) has highlighted the technocratic-scientific period between 1930 and 1995. In the 1970s and 1980s, a new development occurred – there was beginning to be a shift from a strong state-oriented water governance system toward a more open system, in reaction to the growth up of ecological movements and cultural heritage protagonists.

The Netherlands is a good example of the evolution of water management toward more integration and adaptiveness. In response to changing climate conditions, dehydration and soil subsidence, a new paradigm has been adopted that promotes living with water and accommodating water within the built-up space. Water management measures have to be less static and more adaptive to changing climate conditions (Van Stokkom et al. 2005; Meijerink and Dicke 2008). The major drivers of this policy change were the river floods of 1993 and 1995. Some 200,000 people had to be evacuated due to the unexpectedly high levels of discharge of the Dutch rivers. It was acknowledged that the Dutch rivers were not prepared for the consequences of climate change and that more retention capacity was needed in times of high levels of discharge (Warner 2011). In 2000, the various governmental layers agreed on the National Policy Covenant Water. Each river basin and other water basins were obliged to become self-supporting with regard to their water problem. There has been a cautious transition from hierarchy as a coordination mechanism to network coordination (Van Buuren et al. 2010), also strengthened by the European Water Framework Directive of 2000. The dominant position of the water authorities (the Directorate-General for Public Works and Water Management (RWS) and the water boards) has been weakened by the professionalization of interest groups, the expanded role of provinces and municipalities in water management, and the emancipation and activation of the citizen (Wolsink 2006). This paradigm change also changes the way in which water management gains its legitimacy.

In the Dutch water management and flood risk management sector, there has been a cautious transition from quite top-down and expert-led decision-making to network coordination and stakeholder approaches. This is for a large part caused by the professionalization of interest groups and the emancipation and activation of citizens. In this transition not only stakeholder participation took a flight, but rather recently we also witnessed new ways of stakeholder engagement, i.e., stakeholder initiatives or self-organization (Van Buuren et al., 2012a).

2.3 *The Netherlands: Room for the River Program*[3]

While the dominant discourse suggests Room for the River is a simultaneous move from vertical flood defenses to horizontal expansion (widening) of rivers, and from vertical, top-down management to more egalitarian forms of multi-actor network governance, this is by no means a universal interpretation (Warner et al. 2012). In different countries, the concept Space for the River is set in different historical,

[3] This section is largely built upon the overview article by Rijke et al. (2012) and the book on Room for the River program in different countries (Warner et al. 2012).

geographical, and institutional contexts encompassing diversity in physical, political, cultural, historical, and policy factors. These contexts largely determine the way Room for the River takes shape in different countries.[4]

Since 1970s, a transition has been taking place in water management in the Netherlands. In this transition, the traditional sectoral engineering approach to water management is gradually being replaced by an integrated approach incorporating water management, spatial planning, and ecology (van der Brugge et al. 2005). A similar change is taking place in water management in, for example, Europe and North America (Warner et al. 2012). In the 1980s, the technocratic control paradigm was further challenged by the emergence of the concept of "integrated water management" that originally aimed to avoid conflicts between different uses of water resources through improved coordination.

Although these new principles were already applied in the 1970s and the 1980s, it was not until after the 1993 and 1995 near-miss river floods that a new policy window opened for the implementation of integrated water management (van Stokkom et al. 2005). As the river floods each year, the water distributes sediments throughout the floodplain which in turn reduces the space that was initially allowed for annual floods. In 1995, extreme river water levels nearly caused dike breaches and led to the evacuation of 250,000 people and 1 million cattle (Rijke et al. 2012). Dutch national government responded with a new policy line, the Room for the River Directive. This Directive established that water should be considered as a structuring principle for spatial development. It was recommended that mutual adjustment of water and spatial conditions should be the new policy paradigm rather than water as the leading structuring principle (Rijke et al., 2012).

The Room for the River Directive resulted in the approval of the governmental decision for the Room for the River program by the Dutch parliament in December 2006. The total budget was 2.2 million Euro, and the measures should be taken in the period 2006 till 2015. The Room for the River project site encompasses four rivers: the Rhine, the Meuse, the Waal, and the IJssel. The project area is in the Netherlands, but morphological impacts extend upstream into Germany, portions of France, and Belgium and may reach to the Rhine headwaters in Switzerland over time.

The program has a dual objective of (1) improving safety against flooding of riverine areas of the rivers Rhine, Meuse, Waal, IJssel, and Lek by accommodating a discharge capacity of 16,000 m3/s and (2) contributing to the improvement of the spatial quality of the riverine area. At the start of the program, a set of 39 locations was selected for giving more Room for the Rivers through, for example, flood bypasses, excavation of flood plains, dike relocation, and lowering of grayness (Rijke et al. 2012).[5]

---

[4] In France, like in the USA, drives to reduce river degradation due to large dams brought together aquatic ecology, hydrology, and geomorphology. In the Netherlands and Germany, pollution was a more prominent concern, and in the former country, a "hydraulic culture" of damming and diking still dominates thinking (Van Hemert 2008: 108–109) – a heritage that continues to influence interpretations of adaptation in terms of resistance instead of resilience (Warner et al. 2012).

[5] Measures in the plan include placing and moving dikes; depoldering, creating, and increasing the depth of flood channels; reducing the height of the groins; removing obstacles; and constructing a

Although the developments toward integral water management and Room for the River indicated a paradigm shift in Dutch water management, still it is too soon to speak of "deep institutional change" (Verkerk and Van Buuren, 2013; Wolsink 2006). Although the traditional water institutions were opening up to other disciplines, they still were maintaining their power positions. However, the Room for the River program has adopted a new (multilevel) governance approach in which government agencies in different disciplines (e.g., water safety, planning, agriculture, and nature) and at national, regional, and local levels are actively collaborating. The program uses a mix of centralized (national) steering/decentralized (regional) decision-making processes. The decision frameworks for establishing improved water safety and spatial quality are set by the national government, while the plans and designs are formulated and decisions taken by local and regional stakeholders in 39 regional projects. This approach provided the opportunity for decentralized governments to link local issues such as new developments and the development of natural and recreational areas with the water safety agenda (Rijke et al. 2012).

In 2016 most of the 39 projects and measures within the Room for the River program have been completed and implemented. Meanwhile, the Room for the River is considered an "exemplary project" for adopting new governance approaches by the Ministry of Infrastructure and Environment and Rijkswaterstaat.[6]

3 Framework and Methodology

3.1 Theoretical Framework: Citizen Initiative and Government Response

3.1.1 The Rise of Stakeholder-Oriented Approaches

Within a more general trend toward democratic governance, we see all kinds of initiatives to encourage the involvement of stakeholders (citizens, NGOs, etc.) in the decision-making by means of community(-based) or participatory planning and interactive decision-making (Healey 1997, 2007; Leach and Pelkey 2001; Lowndes et al. 2001; Edelenbos 2005; Nye et al. 2011; Edelenbos et al. 2009; Abbas et al. 2014). This tendency to underline the importance of stakeholder engagement in decision-making processes is also visible in current approaches of climate change

"Green River" which would serve as a flood bypass. This will result in lower flood levels. By 2015 the Rhine branches will safely cope with an outlet capacity of 16,000 cubic meters of water per second; the measures implemented to achieve this will also improve the quality of the environment of the river basin.

[6] The lessons learned from the Room for the River program also have international relevance, as the concepts of making space for rivers and new multilevel and interactive governance approaches are also being adopted by other countries (Rijke et al. 2012). For example, the concept of making space for rivers is being applied in countries, such as France, Germany, Hungary, Romania, the UK, and the USA (Warner et al. 2012).

adaptation (Maciejewski Scheer and Höppner 2010; Pahl-Wostl 2006, 2007; Few et al. 2007), adaptive or collaborative water management (Meijerink and Huitema, 2010; Van Buuren 2013), and integrated flood risk management (Green 2002; Marshall et al. 2010; Lupo Stanghellini 2010; Thaler and Priest 2014; Thaler and Levin-Keitel 2015).

Stakeholder engagement means that any group or individual who can affect or is affected by programs, plans, and projects is involved in the decision-making process (Freema, 1984; Lupo Stanghellini 2010). In this view is a more focused approach than public participation as it underlines the deeper more personalized stakes at hand in decision-making (Beierle 2002; Lupo Stanghellini 2010).

Important motives to involve stakeholders in decision-making about water management are (1) gaining increased support by societal actors, (2) improving the quality of decision-making by using their information and solutions, and (3) creating democratic legitimacy and trust by bridging the perceived growing cleavage between citizens and water management agencies (cf Leach and Pelkey 2001; Macpherson 1977; Sorenson and Torfing 2007; Green 2002; Thaler and Levin-Keitel 2015).

However, there have also been concerns with stakeholder engagement in flood risk management. Some authors mention various barriers which make the above-mentioned motives hard to reach (Chess and Purcell 1999). Tseng and Penning-Rowsell (2012) mention different types of barriers related to governmental institutions (lack of institutional support, accountability problems, and closed cultures) but also barriers regarding time, power inequalities, and stakeholder characteristics.

3.1.2 Stakeholder Involvement: Government or Stakeholder Induced?

In literature, there is increasing attention to stakeholder initiative, self-organization, and local self-responsibility community groups (Thaler and Priest 2014; Thaler and Levin-Keitel 2015). Room for stakeholder involvement considerably reduces the government's abilities to directly control the governing process (Sørenson 2002, p. 99). However, even then the government is indirectly ruling, as stakeholder involvement is highly steered and conditioned by public agencies (Abbas et al. 2014). They set the basic objectives and the rules (how to act; how to allocate burdens and benefits) and structures (at what moments in policy-making) for citizen input. The right and opportunity to be involved are not given; it must be continuously negotiated by contesting existing forms of exclusion based on political-administrative choices (Sørenson 2002, p.704).

Rather than operating individually and within the restrictions of government-organized participatory processes, citizens often organize themselves in local groups and take the initiative for collective action (Thaler and Levin-Keitel 2015; Edelenbos and Van Meerkerk, 2016). In such self-organization, citizens, community groups, and social interest groups spontaneously and bottom-up engage in forms of collective action independently from, or in reaction to, government-induced steering processes, structures or decisions, plans, and projects (Edelenbos and Van Meerkerk 2016).

Local stakeholder initiatives often arise from dissatisfaction with the actions of governments and function as a response to proposed government policy that goes beyond plan resistance but is focused on developing alternative plans and projects.

There is thus an important difference between public participation and stakeholder engagement on the one hand and citizen and stakeholder initiatives on the other hand. The former is initiated by the formal policy initiator and structured by rules set by this initiator. It is restricted as far as the initiator deems necessary. The initiator also defines the scope, moments, and methods of participation. The latter, stakeholder initiatives, springs from the self-organizing networks of citizens and community groups who aim to develop their own alternatives, plans, and projects (Edelenbos and Van Meerkerk 2016.

3.2 Research Methodology

In this research we are interested in describing and explaining the developments of citizen initiatives in water management projects in the Netherlands and the responses these initiatives evoke from governmental actors. On the basis of literature about stakeholder involvement in water management and flood risk management, we expect that the goals, resources, and strategies stakeholders employ and the way governmental actors respond explain the role and impact of stakeholder initiatives and plans on decision-making regarding flood risk management. If the goal or strategy of stakeholders is to fight governmental plans for water management, the response by governmental actors is supposed to be negative to stakeholder initiatives and result in lower impacts. If stakeholders have the resources (time, knowledge, expertise, etc.) to develop their plans and initiatives, governmental actors are supposed to be more inclined to positively respond to stakeholder initiatives. This will, in the end, increase the possibility of policy impact.

These expectations and considerations lead to the following variables that form the core of our analytical framework for this exploratory comparative case study:

- *Goals* of local groups: The ambitions of the local groups and the values they want to protect with their initiative.
- Local stakeholder *strategies*: Which strategy or strategies do they apply? These strategies can vary between, for example, reactive (blocking, litigation) or proactive resistance (by developing and selling own ideas).
- Governmental *response*: How do government actors respond to stakeholder strategies? To what extent did the government include the local group in the planning process, leading to processes of co-creation, co-production, and collaboration?
- *Impact*: To what extent did the stakeholder initiative influences the course of events (process) and the content (scope, alternatives analyzed, etc.) of the planning process (Edelenbos et al. 2009; Klijn et al. 2010)? To what degree are the results from the interactive process translated into (new) policy (Knott and Wildavsky 1980; Koontz 2005)?

We conducted a qualitative comparative case study research to explain the role and impact of stakeholder initiatives in water management. This strategy explicitly focuses on comparing cases to find contextual and situational similarities and differences that can be used for explaining the variance in the dependent variable: the impact of a stakeholder initiative on governmental policy (i.e., impact). We used a combination of an instrumental and conventional comparative case study method (c.f. Stake 1998; Yin 1984). We used an instrumental case study approach as we want to find out more about a particular phenomenon, stakeholder initiatives. We explicitly use a conventional case study strategy as our ambition is to gain insights from this case comparison, which in turn may result in further theoretical reflection on stakeholder initiatives in water management.

We conducted the comparative case study research in a focused way structured by the five variables mentioned and operationalized above, to empirically analyze and explain a particular theoretically relevant issue (role and impact of stakeholder initiatives in water management) and generate new insights. This type of research does not (and cannot) yield generalized empirical knowledge about local stakeholder initiatives, but it does provide a detailed and contextualized understanding of how local stakeholders employ their involvement and strategies in flood risk management projects and how governmental actors respond to these strategies.

We have deliberately selected two cases for this qualitative comparative case study research: (1) dike relocation in Lent and (2) a terps plan in Overdiep Polder. These two cases have some core similarities that make case comparison feasible. All cases deal with realizing measures to improve safety from river floods caused by anticipated climate change in the Netherlands. Moreover, all two cases deal explicitly with stakeholder initiatives, in which citizens and societal organizations jointly develop their own plans next to governmental plans for water management.

One case study (Lent) was part of a larger study of adaptive water management in the Netherlands, in which citizen engagement was central (Van Buuren et al. 2010). The second case study, the Overdiep Polder, is based upon an analysis of primary documents and interviews by two of the current authors (see also Winnubst 2011; Roth and Winnubst 2009, 2010; Warner 2011).

The cases also show contextual differences regarding the strategy, resources, and goals of initiators and governmental responses to stakeholder initiatives. Moreover, the cases differ in the dependent impact, the impact of stakeholders on decision-making regarding flood risk management. These differences were not fleshed out beforehand, but were analyzed in more detail during the case comparative research. It is important for comparative case study research that cases have enough differences, in order to gain explanatory power in the comparative analysis.

In all cases in-depth, semi-structured, and open-ended interviews were held. In all cases we interviewed various stakeholders, such as representatives of the Room for the River program, farmers, citizen organizations, municipalities, province representatives, national departments, and nature conservation organizations. Some people (e.g., chair of local interest group, project manager from municipality) were interviewed several times because of their pivotal role in the projects. All interviews were elaborated in transcripts. The transcripts were coded, in which the operationalized core variables were used as main codes. First the data was analyzed per case. The researchers discussed the data per case in different rounds to gain common understanding about the core variables in the study. Second the data was compared between the two cases. Again, all researchers discussed the similarities and differences between the cases in different rounds of discussion in order to reach common ground.

4 Case Studies: Overdiep and Noordwaard[7]

4.1 Introduction of the Cases

In the Dutch water management and flood risk management sector, there has been a cautious transition from quite top-down and expert-led decision-making to network coordination and stakeholder approaches. This is for a large part caused by the professionalization of interest groups and the emancipation and activation of citizens. In this transition not only stakeholder participation took a flight, but rather recently we also witnessed new ways of stakeholder engagement, i.e., stakeholder initiatives or self-organization (Van Buuren et al. 2012a). This is the changing Dutch context of flood risk management in which the two cases took place (Van der Brugge et al. 2005).

The first case, the terps plan in *Overdiep Polder*, is about enlarging discharge capacity in the River Bergsche Maas. The Overdiep Polder (550 ha) is hemmed in between the Oude Maasje and the Bergsche Maas. Confronted with government plans for using the polder for flood storage, the inhabitants of the polder (around 19 households, almost all dairy farmers) developed their own proposal, in cooperation with a water expert. To avoid a long, uncertain planning process, the farmers decided to take the initiative. In their proposal for the polder, the idea was developed to make the polder suitable for flood storage (once in 25 years on average) while keeping its agricultural functions by constructing mounds (or "terps") and rebuilding part of the farms and related property on them. The total number of farming households in the polder had to be reduced from 17 to around 8, ideally by voluntary sales resulting from farmers' decisions to leave the polder. Additional goals were the enhancement of spatial quality and strengthening the agricultural enterprises remaining in the polder.

The second case is the *dike relocation in Lent*. Lent was considered a bottleneck in the river system, as the River Waal between the city of Nijmegen and Lent is very narrow. The municipality of Nijmegen was surprised by the plan for a dike relocation initiated by the national government (Department of Infrastructure and Water Management), because the city had already been given approval by the national government (Ministry of Housing) for a housing project in exactly the same area. In 2002, Nijmegen and the national government signed agreements for a contribution to the costs of a second bridge across the River Waal and compensation for redevel-

[7] The case study sections in this chapter have been grounded in a multiple case study research of three case studies in the Room for the River program. This study will be published in 2017 in *Journal of Environmental Planning and Management* (Edelenbos et al., 2017).

Table 2.2 Which local stakeholders are involved in the projects?

Cases actors	Dike relocation in Lent	Overdiep Polder
Citizen groups	Very much present. Resistance against the breaking down of houses. Different local community groups, individual residents	Inhabitants are mostly farmers. United in one interest group. Very active in "determining their own future"
Farmers	Not present in this case	Individual farmers united in interest group Overdiep Polder
Environmental groups	A few with focus on cultural heritage and nature	Not present
Others	University of Delft; a professor with personal ideas for the development of the area	Habiforum (knowledge institute) province of Brabant
		Steering committee rivers

Source: author

oping the local government's plan for housing. These agreements were based on a preference for the plan for a dike relocation (350 meters inland into the village of Lent). For the proposed dike relocation, 50 houses had to be demolished, which caused much local resistance.

4.2 *Analysis of the Nature of Local Stakeholder Engagement in the Cases*

In this section we systematically analyze the nature of stakeholder engagement in the two cases. We treat subsequently the following aspects: (1) goals of local actors involved in the cases, (2) the strategies the local stakeholders develop and implement, (3) the responses of governmental actors to these strategies, and (4) the impact of stakeholder initiatives (Table 2.2).

4.2.1 Goals of Local Stakeholders

Terps Plan in Overdiep Polder

After an information meeting in May 2000 about government plans for water storage in the polder, four farmers invited a provincial delegate and asked him to be given the opportunity to develop their own alternative plan for combining living, agriculture, and dairy farming with water storage during peak water periods. Farmers' organization ZLTO and the Province of Noord-Brabant supported the farmers in developing their plan. The inhabitants organized into the Overdiep Farmers' Association, which represented them in their negotiations with governmental actors. As only a proportion of the original inhabitants could stay in the polder, several families would have to move out (a process almost finalized at the time of writing this article). Fearing slow procedures, long-lasting uncertainty about the cost-effectiveness of investments and endless delays, most inhabitants shared the wish not to become dependent on externally (government-) developed plans for their polder. As project planning moved on, rifts between the inhabitants (the plausible "stayers" and "leavers") became manifest and increasingly influenced the planning process, the coherence of the Overdiep Farmers' Association, and the relationships between the farmers.

Dike Relocation in Lent

In the dike relocation in Lent case, diverse citizen groups were involved. In reaction to the governmental plan for dike relocation, several local groups were formed. These can be divided into (1) "leavers" (those that have to leave their properties for the dike relocation), (2) "watchers" (those who get a new dike in their front yards), and (3) "stayers" (those who can stay living on the new island). In addition, there were two other local bodies, the village council and an environmental group. The common aim of all these actors was to get the dike relocation off the table.

4.2.2 The Strategies of the Local Stakeholders

Terps Plan in Overdiep Polder

As responsible authority for plan development, the province was in charge of the negotiations with the farmers, including general aspects of the plan like compensation issues. The farmers managed to bring in their alternative ideas. Because there was a strong pressure on the (national and provincial) government to make this showcase project a "success," the farmers, represented by the Overdiep Farmers' Association, had a relatively strong negotiating position in the planning process. They got help from a professional independent process manager who was also very capable in managing media attention.

However, during the planning process, the farmers' group gradually changed into an interest group for "stayers." Hence, in the course of the process, the "leavers" did not have a platform representing their interest. Moreover, the farmers' strategy gradually changed. Depending on their personal and household situation, farmers opted for the best possible solution, whether to start a new farming enterprise in the polder or elsewhere or to stop farming altogether. As a result, the farmers' community partly fell apart, because farmers left their collaborative strategy and opt for a go-alone strategy.

Dike Relocation in Lent

The government plan to relocate the dike 350 meters inland and to demolish 50 houses that caused much resistance among the inhabitants. Their first reaction was to invest positive energy in developing an alternative rather than negative energy in procedural blocking. They estimated that NIMBY behavior would not be a successful strategy, because the government will win in the end. In 2002, the local group developed an alternative, Lentse Warande, in response to the government plan. The Delft emeritus professor (see above) provided the professional support for their plan. The citizens' alternative would make dike relocation and the demolition of dwellings superfluous.

The plan was included in the Environmental Impact Assessment (EIA) procedure, as alternative to the government plan. The local groups got the opportunity to represent citizen interests in the advisory board that was part of the project organization for conducting the EIA studies. Although the advisory board could advice the steering group in decision-making issues, their advice was not-binding.

In the participation procedure of the EIA, more than 500 written "viewpoints" were submitted in support of Lentse Warande, but this did not change the government's decision. After the EIA the project had to follow the national procedure for Room for the River, which includes the possibility for citizens to file complaints and give suggestions. In total 300 letters have been written. In reaction, the permanent committee Infrastructure and Water Management in parliament invited representatives of the Lent Federation to hear their thoughts on this issue.

4.2.3 Governmental Responses

Terps Plan in Overdiep Polder

Here the farmer initiative combined the national objective of making Room for the River with the local interest of strengthening dairy farming. The terps plan will lower the water level during peak discharge in the River Meuse with around 28 cm, which convinced the national government to support the plan. The farmers insisted on getting their plan approved and playing a key role in the process. As a consequence, they were included in the governmental working group, while their expert became an advisor of the executive board, a novelty in the history of water management. Knowing their key position in the planning process, the farmers shared the ownership of the terps plan with the provincial and national governments, as well as the water board that is in charge of implementation. In the Overdiep case, receptivity on the part of the provincial administration was high. However, the key decision-making role on funding remained in the hands of the national government. Sometimes this caused tensions between farmers and province or between the province and national government. These were solved by negotiations, in which the farmers had a relatively strong position (see above).

Dike Relocation in Lent

In the beginning of the project, the city of Nijmegen had signed agreements with the municipality and the national government about the preferred option: dike relocation. In return, Nijmegen was to get compensation for its intended housing plan in the area where the dike had to be moved inland. Nijmegen also negotiated national funding (90 million Euro) for a bridge across River Waal in order to solve congestion problems with the existing bridge. While the alderman of Nijmegen supported the government plan, the city council supported the citizens' alternative. This led to a political schism that also influenced the steering group of the EIA organization. The national government presented acceptance of the government dike relocation plan as a precondition for government funding of the new city bridge. This further diminished the room for maneuver for the alderman and made the citizens feel they were not part of the decision-making and were not taken seriously.

However, under political pressure the aldermen of the municipality decided to involve the citizens in the planning process. They were invited to join a group of stakeholders that could provide non-binding advice. This group advised to involve the plan Lentse Warande in the EIA. Finally, the municipality agreed, leading to more co-production between the civil servants and citizens in making their plan "procedurally and substantially fit" for this assessment, not to really integrate the diversity of ambitions and interests of citizens and the municipality.

4.2.4 Impact

Terps Plan in Overdiep Polder

The impact of the citizens' strategies in the terps plan in Overdiep Polder case can be considered high, as the terps plan was positively received by most inhabitants of the polder and by the other actors involved and adopted. It took a couple of years after its introduction of the terps plan was accepted and adopted by the governmental working group. The citizen initiative developed into a process of co-creation in which representatives of the government (local, regional, and national) collaborated with local stakeholders, a relationship characterized by periods of cooperation and of conflict. In the end the terps plan raised by the inhabitants of the polder became adopted in the governmental implementation strategy.

Dike Relocation in Lent

The impact of the citizens' strategies in the dike relocation in Lent case can be considered "procedural impact" as the citizen's plan, the Lentse Warande, was included in the EIA procedure. In this EIA the Lentse Warande was considered best in its short-term effects. However, the citizens' plan was scoring less positive in long-term climate effects as they were less focused on future climate impacts and considered less the consequences of these impacts for the need for change. For mainly this reason, the option of dike relocation was favored, and the citizens' option was dropped. The citizens fought the outcome even up to the national government, but in the end the dike relocation option was chosen, after which citizen resistance and engagement extinguished.

5 Discussion

In this section we compare and discuss the empirical case study results from Sect. 3. We discuss similarities and differences between the cases regarding the main distinguished variables (goals, resources, strategies, governmental responses, and impact) and critically reflect upon their impact. We also position the gained insights in the current literature and scholarly debate on stakeholder engagement in flood risk management.

The first insight from comparative case analysis is that in both cases, various local stakeholders with different (organizational) backgrounds were engaged in the various flood risk management projects, found each other in developing their own goals (parallel to the governmental plans), and managed to mobilize all kinds of resources to assist them in the formulation and realization of their goals and initiatives. The organized local groups not only used resources like knowledge and information but also social networks and collaborations with knowledge organizations and technical experts. The local groups were well-prepared and devoted much time to mobilizing supportive resources to make their plans and initiatives more substantially sound and convince governmental agencies. The media were often used in airing their complaints about the governmental flood risk management plans and getting exposure for their plans. Other scholars also stress this observation. Thaler and Levin (2015) especially stress the local capacity to act, such as knowledge, time, financial, social, and cultural capital. This local capital and capacity are needed to ensure their interests in flood risk management (see also Kruse and Seidl 2013). This capital and capacity are crucial for realizing impact, as we discuss later in this section.

The second main insight from comparative analysis is the plurality of strategies used by the different stakeholders in the cases. Also some other studies indicate that strong bottom-up community processes are accompanied with strong local leadership and active lobbying to enforce government authorities to cooperate with (local) stakeholders (Thaler and Levin-Keitel 2015). However, our study also reveals that local stakeholders are not just about resisting governmental plans, but they are also much more about developing alternative plans and initiatives to substantiate their voices. It is striking to see that in both cases local stakeholders complemented or combined their defensive strategies with more constructive strategies in which they developed alternative plans that better represented their ideas, interests, and values than the governmental plans. Moreover, stakeholders were actively exploiting their networks in order to expose and realize their plans, including informing the media at crucial moments and getting support from governmental bodies, elected officials, and politicians (lobbying).

The third insight is that, although stakeholder initiatives in the two cases share many similar conditions (like the formal planning procedure as a context, the resistance against the initial public proposal as main driver), they differ considerably in their impact on decision-making about flood risk management plans. In only one case (Overdiep), the impact on decision-making is high because of the adopted plan that is a farmers' initiative and the central role of the residents in the planning process. In the other two cases, the impact is low. In the case of Lent, we see that impact is purely procedural; the stakeholder's plan is incorporated into the formal procedure (EIA), but with no real influence on decision-making.

How can this difference in impact of stakeholder initiatives on decision-making be explained? One aspect is very important, which already has been touched upon in the beginning of this section: empowerment and local capacity. This is also recognized by other scholars (Thaler and Levin-Keitle, 2015; Kruse and Seidl 2013; Kuhlicke et al. 2011) as important assets for local stakeholders to make sure that governmental bodies can't ignore their interests and plans and that they at least have to reconsider their flood risk management strategies. However, as this local capacity was found in both cases, this is necessary but not sufficient to generate impact. Other conditions have thus to be distilled to explain impact.

We found that timing and co-creation are important conditions for developing stakeholder initiatives with impact. In the case of dike relocation in Lent, local input could be integrated in the EIA studies, although local and national governments agreed on compensation measures which locked decision-making on the governmental plan. The timing of the input by stakeholders can be considered bad, but they could not be blamed for this, because local, regional, and national governments had already made deals behind the scenes which made input from other stakeholders difficult. In the case of the Overdiep Polder, the timing of the initiative was better, as governmental plans had not been developed yet and provincial officials were receptive for inputs from local stakeholders. This barrier of timing is also recognized in other studies to stakeholder and public participation (Chess and Purcell 1999). Stakeholders are often only engaged at the later stages of the project cycle, and this negatively influences stakeholders to be involved in the decision-making itself (Tseng and Penning-Rowsell 2012: 256). Decision-making has already proceeded, and the momentum to having real influence and impact is already passed.

Moreover, the nature of the governmental response is important for the chances of impact of stakeholder initiatives. In the case of the Overdiep Polder, a positive response and receptivity to stakeholder initiative were present and evolved into a process of co-creation in which citizens and representatives of lower-level governmental agencies collaboratively worked toward acceptable solutions for flood risk management and spatial planning. In the case of dike relocation in Lent, we see that a process of co-creation was lacking, as the municipal government didn't really have the intention to make the stakeholder initiative a serious "competitor" for their own governmental plan. This factor is found to be very important for the impact of stakeholder initiatives in developing and implementing flood risk management strategies. We can call it the (lack of) institutional susceptibility to bottom-up initiatives. This implies the degree governmental institutions are receptive to initiatives and plans developed by citizens (Tseng and Penning-Rowsell 2012; Edelenbos et al. 2009). Governments don't always approach stakeholder engagement and initiatives in an open and receptive way, but feel threatened by them, as it could imply "hollowing out the state" (Edelenbos 2005). Some even emphasize the political side: introduction of stakeholder engagement in flood risk management has consequences for existing power relations. Some governmental institutions resist this power sharing (Warner 2006, 2011), which causes barriers to integrating local knowledge into their plans (Tseng and Penning-Rowsell 2012: 257). Especially for the rather closed and expert-oriented Dutch flood risk domain, it is difficult to enlarge its susceptibility for outsider initiatives. However, in some case it is present in other's absence, which can be explained by the culture of the governmental institutions and their past experience. In an open culture, public officials are more entrepreneurial and daring to involve input from outside the organization. In a closed culture, public officials are less daring and more defensive to outside input. Also past experience can be relevant. A good past experience, for example, by a pilot, can create a positive attitude by public officials in supporting the citizen initiative. A bad past experience can create reluctant and even negative attitudes and behaviors toward citizen initiatives by public officials.

6 Conclusions

This article has explored the way stakeholders develop alternative flood risk management strategies in the Netherlands. We draw four main conclusions regarding the nature of stakeholder initiatives and the relationship with impact.

First, we can draw the conclusion that in flood risk management, a sector in the Netherlands (but also in other countries) is heavily dominated by governmental agencies and technical expertise and stakeholder initiatives arise beyond the direct control of government (see also Lupo Stanghellini 2010). We see that a "wisdom of the crowds" (Surowiecki 2004) evolves, in which local stakeholders team up with their professional relations (architects, planners, scientists, social community members, etc.) and develop their own plans through a process of self-organization in which they mobilize a wide range of resources and capacities, like media attention, expertise, network relations, finances, etc. (see also Thaler and Priest 2014; Kuhlicke et al. 2011; Kruse and Seidl 2013). Local stakeholders show adaptive behavior in turning their defensive into proactive and productive strategies by developing own alternatives and solutions. This result coincides with observations made by other scholars in the field of climate change and flood risk management (c.f. Nye et al. 2011; Thaler and Levin-Keitel 2015). Our study, however, adds some meaningful insights to this emergence of stakeholder initiatives by relating this changing role of local stakeholders to the impact of stakeholder initiatives on decision-making about flood risk management measures.

Second, our study also shows that local stakeholders use a mix of strategies and resources (media attention, lobbying, constructing own plans, using their networks, etc.) to organize impact. Our study indicates that the exact mix of these strategies helps to explain the impact of the alternative plans developed by local stakeholders. A combination of positive (generating expertise) and negative lobbying (media attention), as well as anchoring the initiative within the formal planning and flood risk management process, seemed to be quite successful. Self-organized initiatives can gain recognition by pressuring (e.g., political lobbying and exposure in the media). However, too much negative lobbying or too much emphasis on airing complaints about governmental actions may become a negative asset, as we saw in the Lent case, as governments can become defensive and opposed to bottom-up initiatives. From our study we can also learn that stakeholder initiatives which are aimed at a goal which is compatible with the public goal are much more easy to realize, compared to initiatives which are aimed to oppose the formal proposal.

Third, another explanation for impact of stakeholder initiatives is that it is important for the viability and impact of self-organization in water management that it is connected as rapidly and effectively as possible to governmental decision-making. This possible connection depends on how government responds to self-organization in flood risk management. This is also stressed in other literatures on participation and self-organization (Margerum 2011; Edelenbos 2005; Feldman and Khademian 2007; Tseng and Penning-Rowsell 2012).). However, we can add to this insight that a positive attitude and response from governmental agencies to stakeholder initiatives may contribute to the explanation of the impact of these initiatives, as our comparative analysis shows that impact of stakeholder initiatives on decision-making increases when governmental agents are more open and responsive to local stakeholder initiatives. This is also influenced by which mix of strategies the initiative is launched (see first conclusion). A positive response followed up by a process of co-creation in which local stakeholders and governmental agencies join hands in further developing the stakeholder plan further increases the chance of impact on decision-making, as the stakeholder initiative gets broader attention, elaboration, and support. For securing impact it is important to combine stakeholder knowledge on the one hand and professional and bureaucratic knowledge on the other, with the possibility that this results in co-produced plans for flood risk management.

Finally, the timing of the stakeholder initiative in the process of water management influences the role and impact of stakeholder initiatives in water management in the two cases. As it has been shown above, we have seen that some substantial promising initiatives come at a bad moment (too late), for example, when governmental agencies have predetermined the preferred measure for flood risk management. Timing and finding the right people in governmental agencies turn out to be crucial factors for generating support for local stakeholder initiatives in flood risk management. Timing is important as this determines the level of receptivity of governmental actors to local stakeholder initiatives. However, the response also depends on to what level the governmental actor really values the input of local stakeholders and creates openings in the decision-making process to incorporate this input.

Water management becomes more and more confronted with stakeholder initiatives (Thaler and Levin-Keitel 2015; Heltberg et al. 2012; Rinaudo and Garin 2005; Petts and Brooks 2006; Sabatier et al. 2005). Therefore, effective and legitimate flood risk strategies increasingly result from local stakeholder initiatives and the interplay of the strategies of local stakeholders and governmental actors. However, this interplay does not emerge spontaneously and needs careful timing and co-creation (Margerum 2011) facing complexity in planning and flood risk management in a productive way (Innes and Booher 2010). Especially from authorities in the flood risk management domain, a more adaptive and receptive stance toward local stakeholder initiatives seems to be necessary to realize flood risk management measures that are not only effective and realized in time but which are also legitimate by doing justice to local values and interests.

The Dutch situation could be sometimes different from the Asian contexts. The Netherlands has a long history of consultation and has strong consensus-seeking culture (Lijphart 1969), which both strongly influence the possibilities to implement interactive governance approaches, specifically citizen initiatives, for example, in the water management domain. In examining Asian cases, it should be noted that in the past (1945–1970), the Netherlands was also characterized by elite politics and hierarchical decision-making. From the 1970s (first generation of participation, see Table 2.1) onward, we see that citizen resisted more and more the elite politics. We also witness that citizens emancipated to well-educated and self-conscious residents, which want more direct influence on decision-making that impacts their daily lives. These processes of resistance and emancipation could be also visible in some Asian countries nowadays.

Interactive governance, with more active citizenry, implies that citizens take more initiative and responsibility in different domains: sustainable energy, health care, and flood risk management. However, pure forms of self-organized citizens, which act with full authority and in splendid isolation, rarely exist in daily (policy) practice. Interactive governance often leads to hybrid forms of governance, in which citizens as well as governmental agencies are active. In this chapter we discussed this by analyzing two case studies in the Room for the River program. We see that an important facilitating factor for realizing impact (effective flood risk management measures) depends on the way governments take a new role that fits the characteristics and nature of the interactive governance processes. In fact, an active co-producing role by governmental agencies assists citizen initiatives to become meaningful and effective (realize impact), for example, by providing resources (expert knowledge, financial means, administrative support, etc.). These resources are needed to develop interactive processes into projects of high impact. This governmental support is oftentimes needed to realize efficient, effective, and legitimate (regarding input, throughput, and output) projects and program. The debate in the Netherlands is about the "expected roles" that governments can take regarding citizen initiatives. In this, one doesn't opt for just one role in all circumstances ("one size fits all"), but for various roles that need to be adjusted to the specific characteristics of the policy domain and the issue (problem) at hand.

In the Netherlands, a distinction is made between four potential roles regarding interactive governance and citizen initiatives: (1) regulatory role, when governments still have a strong legal responsibility; (2) stimulating role, when governments need the input and resources from society in tackling wicked issues (like flood risk management); (3) facilitating role, in case governments take an active role (compare the hands-on meta-governance by Torfing et al. 2012) in assisting citizen initiatives in providing them resources and problem-solving capacity; and (4) letting go role, in which governments provide room (and discretionary authority and power) to citizen initiatives to (largely) work on their own with own responsibilities and accountabilities in solving wicked issues. Each specific circumstance and problem situation has to be assessed in order to see which governmental role is most suitable, applicable, and feasible. So, the situation determines which role governmental agencies are expected to take. This contextual way of looking for the "right roles in the right circumstances" seems a very promising way to find well-functioning co-producing and co-creating relationship between state and non-state actors.

References

Abbas, N.Hl., Irna van der Molen, Manal R. Nader, and Jon C. Lovett. 2014. Citizens' Perceptions of Trust Relationships in the Environmental Management Process in North Lebanon. *Journal of Environmental Planning and Management* 58: 1511. https://doi.org/10.1080/09640568.2014.935757.

Bang, H.P. 2004. Culture Governance: Governing Self-Reflexive Modernity. *Public Administration* 82 (1): 157–190.

Beierle, T. 2002. The Quality of Stakeholder-Based Decisions. *Risk Analysis* 22 (4): 739–749.

Boonstra, B. 2015. *Planning Strategies in an Age of Active Citizenship: A Post-structuralist Agenda for Self-organization in Spatial Planning*. Doctoral thesis. Groningen: PhD Series InPlanning.

Chess, C., and K. Purcell. 1999. Public Participation and the Environment: Do We Know What Works? *Environmental Science & Technology* 33: 2685–2692.

Dalton, R.J. 2008. Citizenship Norms and the Expansion of Political Participation. *Political Studies* 56 (1): 76–98.

De Moor, T. 2013. *Homo Cooperans. Instituties voor collectieve actie en solidaire samenleving, oratie*. Universiteit van Utrecht.

Edelenbos, J. 2005. Institutional Implications of Interactive Governance: Insights from Dutch Practice. *Governance. An International Journal of Policy and Administration* 18 (1): 111–134.

Edelenbos, J., and I.F. van Meerkerk. 2016. *Critical Reflections on Interactive Governance. Self-Organization and Participation in Public Governance*. Cheltenham: Edward Elgar Publishing.

Edelenbos, J., P.J. Klok, and J. van Tatenhove. 2009. The Institutional Embedding of Interactive Policy Making. Insights from a Comparative Research Based on Eight Interactive Projects in the Netherlands. *American Review of Public Administration* 39 (2): 125–148.

Edelenbos, J., I. van Meerkerk, and T. Schenk. 2016. *The Evolution of Community Self-organization in Interaction with Government Institutions: Cross-Case Insights from Three Countries*. American Review of Public Administration. http://arp.sagepub.com/content/early/2016/06/14/0275074016651142.full.pdf

Edelenbos, J., M.W. van Buuren, D. Roth and M. Winnubst (2017), Stakeholder Initiatives in Flood Risk Management: Exploring the Role and Impact of Bottom-Up Initiatives in Three 'Room for the River' Projects in the Netherlands, Journal of Environmental Planning and Management. 60 (1), pp. 47–66 Accepted for publication

Feldman, M.S., and A.M. Khademian. 2007. The Role of the Public Manager in Inclusion: Creating Communities of Participation. *Governance* 20 (2): 305–324.

Few, R., K. Brown, and E.L. Tompkins. 2007. Public Participation and Climate Change Adaptation: Avoiding the Illusion of Inclusion. *Climate Policy* 7 (1): 46–59.

Freeman, R.E. 1984. *Strategic Management: A Stakeholder Approach*. Boston: Pitman.

Green C. H. 2002. Flood Management from the Perspective of Integrated Water Resource Management. Paper given at the *2nd International Symposium on Flood Defence*, Beijing.

Healey, P., 1997. *Collaborative Planning: Shaping Places in Fragmented Societies* (2nd Rev. ed. by UBC press in 2006), Palmgrave, Hampshire.

———. 2007. *Urban Complexity and Spatial Strategies: Towards a Relational Planning for Our Times*. London: Routledge.

Heltberg, R., H. Gitay, and R.G. Prabhu. 2012. Community-Based Adaptation: Lessons from a Grant Competition. *Climate Policy* 12 (2): 143–163.

Innes, J.E., and D.E. Booher. 2010. *Planning with Complexity. An Introduction to Collaborative Rationality in Public Policy*. London: Routledge.

Irvin, R.A., and J. Stansbury. 2004. Citizen Participation in Decision Making: Is It Worth the Effort? *Public Administration Review* 64 (1): 55–65.

Klijn, E.H., J. Edelenbos, and A.J. Steijn. 2010. Trust in Governance Networks: Its Impact and Outcomes. *Administration and Society* 42 (2): 193–221.

Knott, J., and A. Wildavsky. 1980. If Dissemination Is the Solution, What Is the Problem? *Knowledge: Creation, Diffusion, Utilization* 1 (4): 537–578.

Kooiman, J., ed. 1993. *Modern Governance: New Government–Society Interactions*. London: Sage.

Koontz, T. 2005. We Finished the Plan, So Now What? Impacts of Collaborative Stakeholder Participation on Land Use Policy. *The Policy Studies Journal* 33: 459–481.

Kruse, S., and L. Seidl. 2013. Social Capacity for Drought Risk Management in Switzerland. *Natural Hazards and Earth System Sciences* 13 (12): 3429–3441.

Kuhlicke, C., et al. 2011. Perspectives on Social Capacity Building for Natural Hazards: Outlining an Emerging Field of Research and Practice in Europe. *Environmental Science & Policy* 2011: 804–814.

Leach, W., and N. Pelkey. 2001. Making Watershed Partnerships Work: A Review of the Empirical Literature. *Journal of Water Resources, Planning and Management* 127 (6): 378–385.

Lijphart, Arend. 1969. Consociational Democracy. *World politics* 21 (02): 207–225.

Lijphart, Arend, and Don Aitkin. 1994. *Electoral Systems and Party Systems: A Study of Twenty-Seven Democracies, 1945–1990*. Oxford: Oxford University Press.

Lintsen, H. 2002. Two Centuries of Central Water Management in the Netherlands. *Technology and Culture* 43: 549–568.

Lowndes, V., L. Pratchett, and G. Stoker. 2001. Trends in Public Participation: Part 2 – Citizens's Perspectives. *Public Administration* 79 (2): 445–455.

Lupo Stanghellini, P.S. 2010. Stakeholder Involvement in Water Management: The Role of the Stakeholder Analysis Within Participatory Processes. *Water Policy* 12: 675–694.

Maciejewski Scheer, A., and C. Höppner. 2010. The Public Consultation to the UK Climate Change Act 2008: A Critical Analysis. *Climate Policy* 10 (3): 261–276.

MacPherson, C.B. 1977. *The Life and Times of Liberal Democracy*. Oxford: Oxford University Press.

Margerum, R.D. 2011. *Beyond Consensus: Improving Collaborative Planning and Management*. Boston: MIT Press.

Marien, S., M. Hooghe, and E. Quintelier. 2010. Inequalities in Non-institutionalised Forms of Political Participation: A Multi-level Analysis of 25 Countries. *Political Studies* 58 (1): 187–213.

Marshall, K., K.L. Blackstock, and J. Dunglinson. 2010. A Contextual Framework for Understanding Good Practice in Integrated Catchment Management. *Journal of Environmental Planning and Management* 53 (1): 63–89.

Mayer, I., J. Edelenbos, et al. 2005. Interactive Policy Development: Undermining or Sustaining Democracy? *Public Administration* 83 (1): 179–199.

Meijerink, S., and W. Dicke. 2008. Shifts in the Public-Private Divide in Flood in Management. *International Journal of Water Resources Development* 24 (4): 499–512.

Meijerink, S., and D. Huitema. 2010. *Water Transitions, Policy Entrepreneurs and Change Strategies: Lessons Learned Water Policy Entrepreneurs: A Research Companion to Water Transitions Around the Globe*. Cheltenham: Edward Elgar Publishing.

Nye, M., S. Tapsell, and C. Twigger-Ross. 2011. New Social Directions in UK Flood Risk Management: Moving Towards Flood Risk Citizenship. *Journal of Flood Risk Management* 4 (4): 288–297.

Pahl-Wostl, C., 2006. The Importance of Social Learning in Restoring the Multifunctionality of Rivers and Floodplains. *Ecology and Society*, 11 (1), art. 10. http://www.ecologyandsociety.org/vol11/iss1/art10/

———. 2007. Transitions Towards Adaptive Management of Water Facing Climate and Global Change. *Water Resources Management* 21 (1): 49–62.

Petts, J., and C. Brooks. 2006. Expert Conceptualizations of the Role of Lay Knowledge in Environmental Decision-Making: Challenges for Deliberative Democracy. *Environment and Planning A* 38: 1045–1059.

Rhodes, R.A.W. 1997. *Understanding Governance. Policy Networks, Governance, Reflexivity, and Accountability*. Buckingham: Open University Press.

Rijke, J., S. van Herk, C. Zevenbergen, and R. Ashley. 2012. Room for the River: Delivering Integrated River Basin Management in the Netherlands. *International Journal of River Basin Management* 10 (4): 369–382.

Rinaudo, J.D., and P. Garin. 2005. The Benefits of Combining Lay and Expert Input for Water-Management Planning at the Watershed Level. *Water Policy* 7: 279–293.

Roth, D., and M. Winnubst. 2009. Reconstructing the Polder: Negotiating Property Rights and "Blue Functions" for Land. *International Journal of Agricultural Resources, Governance and Ecology* 8 (1): 37–56.

———. 2010. Overdieps polderen. Enkele aspecten van de planvorming voor de Overdiepse Polder. *Bestuurskunde* 2010 (3): 52–64.

Sabatier, P.A., W. Focht, M. Lubell, Z. Trachtenberg, A. Vedlitz, and M. Matlock, eds. 2005. *Swimming Upstream. Collaborative Approaches to Watershed Management*. Cambridge, MA: MIT Press.

Sørenson, E. 2002. Democratic Theory and Network Governance. *Administrative Theory and Praxis* 24 (4): 693–720.

Sørenson, E., and J. Torfing. 2007. *Theories of Democratic Network Governance*. New York: Palgrave Macmillan.

Stake, R.E. 1998. Case Studies. In *Strategies of Qualitative Inquiry*, ed. N.K. Denzin and Y.S. Lincoln, 86–109. Thousand Oaks: Sage.

Surowiecki, J. 2004. *The Wisdom of Crowds: Why the Many Are Smarter Than the Few and How Collective Wisdom Shapes Business, Economies, Societies and Nations*. New York: Doubleday.

Teisman, G.R., M.W. van Buuren, and L.G. Gerrits. 2009. *Managing Complex Governance Systems*. London: Routledge.

Thaler, T., and M. Levin-Keitel. 2015. Multi-level Stakeholder Engagement in Flood Risk Management – A Question of Roles and Power: Lessons from England. *Environmental Science and Policy* 55: 292–301.

Thaler, T., and S. Priest. 2014. Partnership Funding in Flood Risk Management: New Localism Debate and Policy in England. *Area* 46 (4): 418–425.

Torfing, J., B.G. Peters, J. Pierre, and E. Sørensen. 2012. *Interactive Governance. Advancing the Paradigm*. Oxford: Oxford University Press.

Tseng, C.-P., and E.C. Penning-Rowsell. 2012. Micro-political and Related Barriers to Stakeholder Engagement in Flood Risk Management. *The Geographical Journal* 178 (3): 253–269.

Van Buuren, M.W. 2013. Knowledge for Water Governance. Trends, Limits, and Challenges. *International Journal of Water Governance* 1 (1–2): 157–175.

Van Buuren, M.W., J. Edelenbos, and E.H. Klijn. 2010. *Gebiedsontwikkeling in woelig water. Over water governance bewegend tussen adaptief waterbeheer en ruimtelijke besluitvorming*. Den Haag: Lemma.

Van Buuren, M.W., E.H. Klijn, and J. Edelenbos. 2012a. Democratic Legitimacy of New Forms of Water Management in the Netherlands. *International Journal of Water Resources Development* 28 (4): 629–645.

Van Buuren, M.W., J. Edelenbos, and J.F. Warner. 2012b. Space for the River: Governance Challenges and Lessons. In *Making Space for the River. Governance Experiences with Multifunctional River Flood Management in the US and Europe*, ed. J.F. Warner, M.W. van Buuren, and J. Edelenbos, 187–201. London: IWA Publishing.

Van Meerkerk, I., B. Boonstra, and J. Edelenbos. 2013. Self-Organization in Urban Regeneration: A Two-Case Comparative Research. *European Planning Studies* 21 (10): 1630–1652.

van Stokkom, H.T.C., A.J.M. Smits, and R.S.E.W. Leuven. 2005. Flood Defense in the Netherlands; A New Era, A New Approach. *Water International* 30 (1): 76–87.

Warner, J. 2006. More Sustainable Participation? Multi-stakeholder Platforms for Integrated Catchment Management. *International Journal of Water Resources Development* 22 (1): 15–35.

———. 2011. *Flood Planning. The Politics of River Interventions*. London: IB Tauris.

Warner, J.F., M.W. van Buuren, and J. Edelenbos, eds. 2012. *Making Space for the River. Governance Experiences with Multifunctional River Flood Management in the US and Europe*. London: IWA Publishing.

Winnubst, M., 2011. Turbulent Waters; Cross-Scale Conflict and Collaboration in River Landscape Planning. Ph.D., Nijmegen: Radboud University Nijmegen.

Wolsink, M. 2006. River Basin Approach and Integrated Water Management: Governance Pitfalls for the Dutch Space-Water-Adjustment Management Principle. *Geoforum* 34 (4): 473–487.

Yin, R.K. 1984. *Case Study Research: Design and Methods*. Beverly Hills: Sage.

Jurian Edelenbos is Professor of Interactive Governance in the Department of Public Administration and Sociology, Erasmus School of Social and Behavioral Sciences, Erasmus University Rotterdam, the Netherlands. He is also Academic Director of the Institute for Housing and Urban Development Studies and Director of the International PhD Program Urban Governance and Development. He lectures and conducts research on community-based initiatives, participation, trust, interorganizational collaboration, and boundary spanning leadership and legitimacy in the fields of urban development, water governance, and sustainable energy.

Chapter 3
Contextual Factors Affecting the Modes of Interaction in Governance: The Case of Dam Removal in Japan

Tomohiko Ohno

Abstract This chapter discusses the first Japanese case of dam removal policy. The Arase Dam in Kumamoto Prefecture was built for hydropower generation as a symbol of economic development in 1955. After its construction, local residents came to realize the damage caused by the dam, such as flood damage and eutrophication of the reservoir. In parallel with the anti-dam construction movement, at another site in the same river basin, the local village and fishermen began to ask the prefectural government to remove the Arase Dam when the permit for hydropower generation expired in 2003. Until the governor of the prefecture decided to remove the dam, there had been a series of complex interactions among different levels of government, local residents, fishery cooperatives, and downstream farmers, with political dynamics from changes of top leaders in elections. This chapter focuses on the contextual factors in interactive governance and stresses the importance of resistance strategy, which has been discussed in case studies of local commons in Japan, rather than collaborative governance discussed in many water governance studies.

Keywords Dam removal · Arase Dam · Interactive governance · Collaborative governance · Resistance strategy · Power imbalance · Contextual factor · River policy · Japan

1 Introduction

The concept of governance has attracted scholarly attention over the last few decades. The relatively new concept of "interactive governance" has emerged as an alternative to traditional government practices (Torfing et al. 2012). Dating back several decades, researchers have noted the decline of trust in traditional government agencies among industrialized societies (Crozier et al. 1975) because they have difficulty in dealing with new and complex social issues, such as environmental

T. Ohno (✉)
Faculty of Law, Institute of Human and Social Sciences, Kanazawa University, Kanazawa, Japan
e-mail: t.ohno@staff.kanazawa-u.ac.jp

K. Otsuka (ed.), *Interactive Approaches to Water Governance in Asia*,
https://doi.org/10.1007/978-981-13-2399-7_3

protection. Hence, much attention has been paid to collaborative approaches toward natural resource management (e.g., Koontz 2004; Sabatier et al. 2005).

While the new types of governance have been studied in various countries, the applicability of those concepts has to be examined carefully in various social, economic, and legal settings. As democracy varies among countries and regions (Lijphart 1999), we can assume that the mode of governance also varies among countries and regions.

In this chapter, we examine the applicability of interactive governance as an emerging concept in the Japanese watershed governance issue. While interactive governance has been introduced in several studies in Japan (Hori 2011), it has not been applied empirically in Japanese cases. Watershed governance is one of the relevant issues to examine as it is a "wicked" problem (Rittel and Webber 1973) to govern. There are diverse stakeholders with occasionally countervailing interests in watershed governance. Complexity and scientific uncertainty of a watershed ecosystem also makes it difficult to deal with governance issues (Ohno 2013).

In 1997, Japan revised some fundamental legal structures for river administration to expand the scope of public involvement in planning stages of each river's fundamental management plan. Prior to this revision, the planning process had been almost exclusively under bureaucratic control, although several river-control works significantly influenced livelihoods within the targeted watersheds. Such government-centered governance often created severe disputes over river engineering works, such as dam construction, discharge channel works, and weir constructions. To cope with the widespread criticism over the river administration by the existing government, the River Law was revised in 1997. Another remarkable revision of the River Law in 1997 saw the inclusion of "environmental conservation" into the policy objectives for river administration. Those legal changes were positively evaluated with hopes for the transition to sustainable watershed governance. Some best practices of collaborative and interactive planning at the watershed have been reported (e.g., Obitani 2003); however, the realities of watershed governance after the legal reform have been critically examined (Ohno 2013). Some dams are still in dispute and are going to be built despite the strong opposition by inhabitants (Seki et al. 2015). The linkage between existing governance and emerging interactive governance is the topic to be studied further in the Japanese watershed governance context.

An interesting case in this regard is the removal of the Arase Dam, widely known as the first instance of large-scale dam removal in Japan. Residents near the site originally suggested removing the dam, and the prefectural government decided to proceed with the same in 2003. However, a newly elected prefectural governor reversed this decision, giving rise to a fierce campaign for its removal. Eventually, the governor changed his position, and the removal work was completed in March 2018. The case study of those complicated interactions between residents and government organizations will offer beneficial insights for the future directions of watershed governance in Japan.

Although studies on the dam removal would provide useful insights to governance literature, these studies in the context of governance studies are limited at present. While a series of studies on the dam removal by political scientists (Lowry 2003, 2005, 2009; Mertha and Lowry 2006) focus on the politics and policy process leading to the dam removal, the theoretical contribution to governance literature has not been discussed clearly. With only a few exceptions, including Lowry's works, studies on dam removal have been conducted mostly in the natural sciences. This tendency applies to studies of the Arase Dam as well, which have included a geochemical assessment of how the dam removal impacted the mouth of the river and the adjoining sea (Young and Ishiga 2014) and overall reports on environmental changes anticipated after the removal (Tsuru 2013). Only Abe (2007) has studied this case from a social science viewpoint, discussing the history of the dam and its surrounding communities along with social movements advocating for its removal. However, Abe's analysis covers only the period up to 2005 and does not discuss the subsequent policy process that featured a withdrawal of the first decision to dismantle the dam and then the final decision to remove it. Despite the theoretical concern for understanding the linkage between the existing government and the emerging interactive governance in the Japanese watershed governance context, the processes and interactions that led to the decision to remove the Arase Dam have not previously been studied.

The subsequent sections are as follows. We first review the literature on interactive governance in Europe and environmental or resource governance in Japan. Since we can trace back the conceptual roots of interactive governance in Europe, we first review the literature in Europe. Considering the context-dependent nature of governance, surveying the literature not only in Europe but also Japan will be significant. Furthermore, the basic legal framework of watershed governance will be briefly summarized. After explaining the methods and data used in this study, the historical transition of the governance of Arase Dam will be described comprehensively. We will discuss the characteristics of watershed governance for the dam removal and the influences of contextual factors that determine the mode of governance.

2 Literature Review

2.1 Interactive Governance Literature in European Countries

Interactive governance as a new concept has been discussed mainly by scholars in EU countries. Kooiman (1993), a seminal work opening the field of governance literature, refers to the interactive features of new emerging governance. Kooiman (2003) also suggested that the interactions in the study of governance should be emphasized more. Kooiman used the term "interaction" to refer to "a mutually influencing relation between two or more actors or entities" (Kooiman 2003).

While several researchers define interactive governance in slightly different manners, the common denominator is the social problem-solving process with divergent actors. For instance, Kooiman et al. (2005) define interactive governance as "the whole of interactions taken to solve societal problems and to create societal opportunities; including the formulation and application of principles guiding those interactions and care for institutions that enable and control them." In the comparative analysis of Dutch local governments, Edelenbos (2005) defines interactive governance as "a way of conducting policies whereby a government involves its citizens, social organizations, enterprises, and other stakeholders in the early stages of public policy making." As an introductory chapter of the encompassing book on interactive governance, Torfing et al. (2012) refer to interactive governance as "the complex process through which a plurality of actors with diverging interests interact in order to formulate, promote, and achieve common objectives by means of mobilizing, exchanging, and deploying a range of ideas, rules, and resources."

While the interactive governance literature offers several guiding principles, room remains for further scholarly development. First, the empirical application has been limited to case studies in European countries, such as Dutch regional development studies (Edelenbos 2005; Edelenbos et al. 2010). Fishery governance is a field where scholars have attempted to apply interactive governance theory in empirical investigation. Kooiman et al. (2005) and Jentoft and Bavinck (2014) mostly discuss the global trend of fisheries and the legal framework for fisheries in general. Their focus is not limited to specific countries; nevertheless, most of them are European scholars. Examining the interactive governance studies in a different political, cultural, and economic context will add a new insight into the past studies. Second, the similarities and differences between interactive governance and other new forms of governance are not clear enough. Some scholars discuss the similar forms of governance with interactive governance as "collaborative governance" (Ansell and Gash 2007) or "participatory governance" (Fisher 2012). A clarification of those new modes of governance is needed for conceptual development and would be a beneficial contribution to the governance literature as a whole. As interactive governance studies have been conducted in a relatively limited region, empirical investigation in other social and political contests would be a relevant research strategy.

2.2 *Environmental and Resource Governance Literature in Japan*

Governance has been a significant research topic in various academic fields also in Japan for a few decades. As the contribution to environmental studies from political scientists is relatively limited to Japan, environmental and resource governance has been studied by economists, sociologists, anthropologists, and so on.

One of the topics we should pay attention to among those studies is the collaborative governance of natural resources, especially those as the local commons. Inoue (2004) proposes the collaborative governance of the tropical forest in Indonesia with the collaboration of diverse stakeholders including indigenous resource users, local government, business enterprises, international NGOs, and global citizens.[1]

[1] Inoue (2004) also discusses that each stakeholder should have differentiated legitimacy according to their dependence on the targeted resource.

Mitsumata and Saitoh (2010), Miwa and Mitsumata (2010), and Mitsumata (2013) complement Inoue's concept of collaborative governance by emphasizing the possible adversarial effects outside the local community governing the natural resources in their own community. They discuss that a collaborative relation over the governance of natural resources is sometimes quite difficult to achieve in the case of an unfillable power imbalance between stakeholders. For instance, as briefly mentioned in the introductory section, several river development projects like dam constructions have been advanced by government agencies despite strong opposition and protest activities by inhabitants who receive negative effects from the projects. Many local commons under the external pressure for privatization and nationalization experienced their demise. As collaboration is difficult to realize in those situations, a series of Mitsumata and his colleague's articles proposes "resistance strategy" to complement collaborative governance strategy. They refer to resistance strategy as "the strategy that members of local commons save their own commons by earning the support from various entities, appealing their legitimacy, and protesting the outsiders who (1) have a precise intention to degrade or demise local commons or (2) unintentionally lead to the collapse of prerequisite for maintaining the commons" (Mitsumata and Saitoh 2010). In the case where prerequisites for a collaborative relation are not met due to external or adverse impacts to local commons, the resistance strategy would be possible for the members of the commons to protect their own resources and livelihoods. While the applicability of a resistance strategy has been examined in the case of property wards' resistant responses to local government entities to maintain their autonomy over own resources and institutions (Mitsumata and Saitoh 2010), few studies develop their arguments.

We can draw implications from those studies on governance strategies that collaboration is not the sole answer but one of the eligible modes of environmental governance. That eligibility is difficult to assess in general; however, it is certain that we should take its surrounding context as a prerequisite for collaboration into consideration.

3 Basic Legal Framework for Water Governance in Japan

In the Japanese legal system, there is no law that encompasses watershed governance as a whole. The River Law set the fundamental structure of river governance, although the scope is not an entire watershed area but limited only to the area within the river. Most of the rivers in Japan are subject to the River Law.

Under the River Law provision, government entities are generally designated as river administrators responsible for managing each river. The level of government entities differs from municipal government to national government according to the classes of rivers.[2] In case a river is regarded as very important socially and economically, it is designated as Class A. A river with moderate social and economic

[2] Japan has the two-tier local government system, including prefectures and municipalities. Municipalities include cities (Shi), towns (Cho), and villages (Son or Mura). For the details on Japanese local government, see Reed (1986) and Jacobs (2003).

importance is designated as Class B. Among the rest of the rivers, municipal mayors can designate secondary rivers if the need arises. The Ministry of Land, Infrastructure, Transport and Tourism (MLIT), a national government entity, is designated as a river administrator of Class A rivers (Article 9); the prefectural governor is authorized as a river administrator of Class B rivers (Article 10); and a municipal mayor is chosen as a river administrator of secondary rivers (Article 100). Those provisions on river administrators are considered to entitle encompassing power to government organizations in Japanese river governance (Miyoshi 2007).

With regard to water resource use, there are two kinds of water use rights in Japan. One of them is the "licensed water use right," which needs to be granted by the river administrator on the basis of the amount of water withdrawn from the river. These rights have been established according to the Article 23 of the River Law in 1896. These rights are relatively new as compared to "customary water use rights" as discussed below and often used for industry or urban water supply. As this right is a permission or license rather than property right, river administrators turn out to be influential in determining permissions and renewals of water use licenses.

The other one is "customary water use right," which has its legal basis on the Article 11 of Ordinance for the River Law in 1986. These rights are based on the customary rules for water use within or between village communities and mainly used for irrigation. This right is established to approve existing water use rules prior to the River Law in 1986 and is more a private property issue than a licensed water use right.

4 Method

Since governance in this case is relatively complex and has not previously been comprehensively studied, we adopted the process tracing method (Beach and Pedersen 2013), which involves describing the details of an event from related documents and interviews. To gain an overall understanding of the case, we collected newspaper articles comprising detailed information on the policy process and conducted interviews of key stakeholders.

4.1 Data Collection

Using five major newspaper databases, including Kumamoto Nichinichi Shinbun, Asahi Shinbun, Mainichi Shinbun, Yomiuri Shinbun, and Nikkei Shinbun,[3] we identified articles containing the key phrase "Arase Dam." To check the reliability of the information presented in these articles, we also visited the dam and the surrounding area, where we confirmed the present situation of the river environment and interviewed several key persons in the dam removal campaign. Interviews were

[3]Among those newspapers, only Kumamoto Nichinichi Shinbun is the local newspaper published at Kumamoto; the others are the nationwide newspapers.

conducted since 2013 with a total of 23 persons: 9 fishermen, 4 persons engaged in forestry, and 10 inhabitants along the Arase Dam site. Additionally, we collected related official documents including the recoded minutes of the Kumamoto prefectural parliament. We cross-checked the data collected so as to confirm the validity of the information.

4.2 Overview of the Targeted Area

The Arase Dam is located in the midstream of the Kuma River, Kumamoto Prefecture, Japan (Fig. 3.1). The Kuma River extends for 115 km and has a watershed area of 1880 km^2, or almost 25% of the area of Kumamoto Prefecture. The Kuma River is designated as a Class A river. The total population living within the watershed is about 250,000 now, but it peaked at around 350,000 during the 1950s and 1960s (Fig. 3.2). The Kuma River is famous for its yield of sweetfish [*Plecoglossus altivelis*]. Figure 3.3 illustrates the percentage of workers engaged in primary industries, such as fisheries, forestry, and agriculture, in each municipality in the watershed. The average percentage employed in the primary industries of these municipalities is 20.9%, which is higher than both the national average of 5.1% and the prefectural average of 10.5%. Although the percentage in Hitoyoshi is much lower than in other municipalities in the watershed, Hitoyoshi has a higher percentage of tertiary industry workers and is famous for its tourism industry, including hot springs and riverboat recreation, which depend largely on ecosystem services from the Kuma River. We often observe the uneven rate of workers in

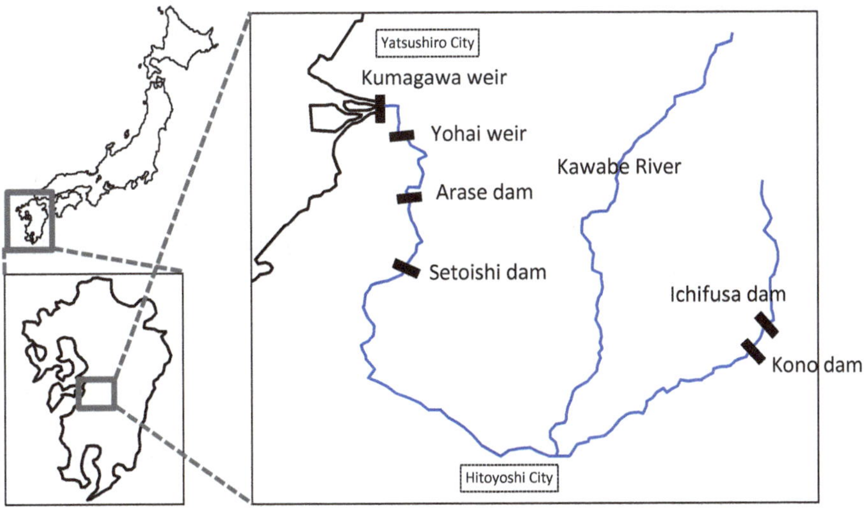

Fig. 3.1 Kuma River watershed map
Source: Compiled by the author based on the map provided by the Ministry of Land, Infrastructure, Transport and Tourism

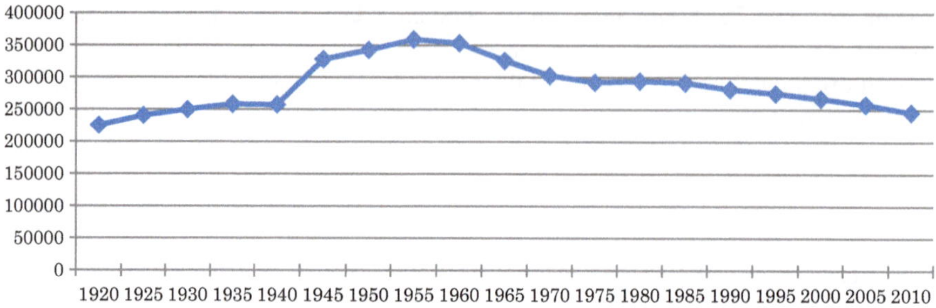

Fig. 3.2 Population changes in the Kuma River watershed area, 1920–2010
Note: Total population is calculated by adding the populations of Yatsushiro City, Hitoyoshi City, Ashikita Town, Nishiki Town, Taragi Town, Yunomae Town, Mizukami Village, Sagara Village, Itsuki Village, Yamae Village, Kuma Village, and Asagiri Town. The data source for each municipality's population is the Population Census for each year
Source: Compiled by the author based on the data

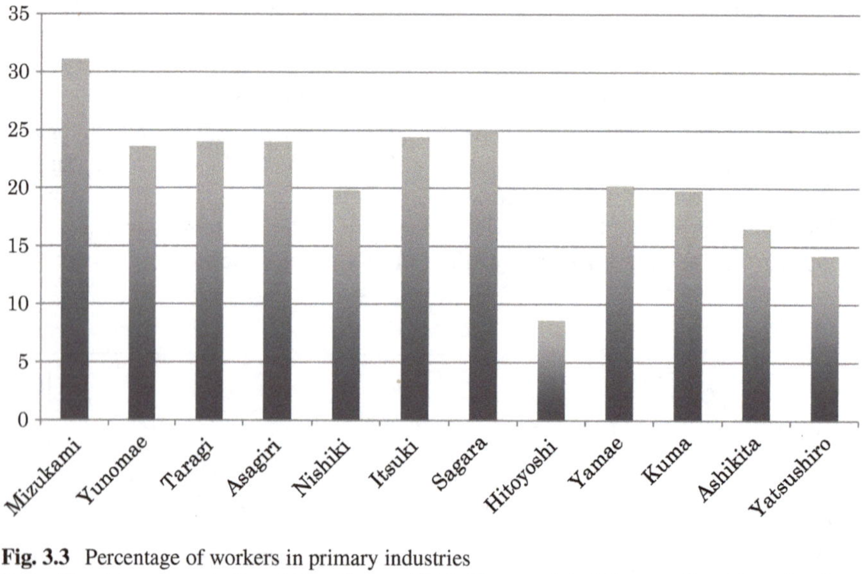

Fig. 3.3 Percentage of workers in primary industries
Note: The data source is Population Census 2010. The municipalities are listed in order from upstream (left) to downstream (right)
Source: Compiled by the author based on the data

primary industries between upstream (high rate) and downstream (low rate); however, those statistics indicate that people's livelihoods, including those at midstream and downstream areas, rely more heavily on the river ecosystem in this watershed than in most other areas.

The Kuma River General Development Project was initiated in 1951. Four large dams, including the Arase Dam, were constructed along the Kuma River for hydropower generation or flood control. The Kawabegawa Dam, which was to be installed on the Kawabe River, a branch of the Kuma River, attracted nationwide attention because of the campaign against its construction.

The Arase Dam was built in 1955 for hydropower generation at Sakamoto Village, which merged with Yatsushiro City in 2005. Arase is a concrete gravity dam,[4] 25 meters high and 210 m wide. The dam's operating body is the Kumamoto Bureau of Enterprise, a branch of the Kumamoto prefectural government responsible for public utilities.

5 Results[5]

5.1 Dam as a Symbol of "Development" (1960s)

During the construction of the Arase Dam in the 1950s, Kumamoto Prefecture was suffering from a shortage of electricity, especially for industrial use. Hydropower was the main source of electricity then. Newspaper articles reported that power outages frequently occurred in Kumamoto,[6] and their electricity supply depended entirely on the amount of rainfall.[7] To deal with the electricity supply problem, Kuma River, known as its abundant river flow, received remarkable prefectural-wide attention. Governor Sakurai decided to start a regional economic development project at Kuma River watershed with reference to the experiences of the TVA (Tennessee Valley Authority) project in the United States. His basic idea of the development project as a "small TVA" was to install several hydropower generation plants and enable invitations from industrial factories in that area.[8] This project was entitled the "Kuma River General Development" project. Its first step was the construction of the Arase Dam.

The building of the Arase Dam was viewed as a symbol of "development" like other dams built in the same period. The prefectural government made a documentary film about the Arase Dam construction,[9] portraying the prefectural-wide expectation for industrial promotion and the magnificence of construction works, and included a lavish ceremony for celebrating the completion of the dam construction.

A publication on local history (Sakamoto Village History Editorial Committee 1990) recorded that the dam project was painful for those who were forced to move away from their long-established residences as the construction work rapidly completed with the help of inhabitants around the dam site. The Arase Dam was

[4] It is made from concrete and is "called a gravity dam because gravity holds it down to the ground stopping the water in the reservoir pushing it over" (The British Dam Society 2010).

[5] See Appendix for the overall process of Arase Dam construction and removal from the 1950s to 2010.

[6] *Kumamoto Nichinichi Shimbun*, May 25, 1951.

[7] *Kumamoto Nichinichi Shimbun*, February 11, 1951.

[8] *Kumamoto Nichinichi Shimbun*, December 23, 1954.

[9] This film is available at the following website (http://www.kagakueizo.org/create/other/5533/).

built in merely 22 months even though the project influenced its surrounding communities in various ways, such as by the relocation of 119 households, compensation for fisheries and log rafts, and new road construction as an alternative to the former use of river transportation. We cannot find any record of fierce opposition to the construction of the dam then; its completion was celebrated throughout the prefecture as a sign of the area's economic development.

5.2 *Dam as a Source of "Nuisance" (1970s–1990s)*

Since the construction of the Arase Dam, local residents have reported various types of damage attributable to the dam. The first trigger of change in the inhabitants' attitude toward Arase Dam was the flood damage caused by heavy rainfall in 1965. A resident along the Kuma River told that the flood in 1965 was entirely different from the previous ones in terms of its "quality." According to him, the previous flood flow was clean and gradually increasing; however, it turned muddy containing the sludge in the reservoir and rapidly increasing due to Arase Dam.[10]

This flood was critically reported in local news published by the community center in the Sakamoto Village. It described the flood damage as a consequence of the inappropriate operations of the existing three dams at Kuma River, including Arase Dam. According to the local news article, flood damage worsened after the construction of the dam, and, as a result, inhabitants within the watershed were suffering.

Inhabitants along the dam reservoir were also experiencing flood damage due to the rising water level of the reservoir. The water level rise during floods was considered to be a result of the accumulation of sands and soils at the dam reservoir. It caused flood damage to the housing area around the reservoir where no such damage had previously occurred. The grounds of some housing areas around the reservoir were raised up by embankment works to cope with the new flood threats.[11]

Inhabitants along the Arase Dam organized an association to advocate the flood damages caused by the dam and negotiate with prefectural government being the administrator of the dam for a compensation of those damages. Nevertheless, their activities did not come to fruition due to the political pressures placed on the association.

Moreover, it became apparent that Arase Dam had harmful impacts on the ecosystem around the dam site. The dam's disruption to water flow caused a decrease in the number of migratory species of fish, like eel and sweetfish, which were well known for their abundance in the Kuma River.[12] The Ministry of Construction

---

[10] Interview, March 6, 2016.

[11] Details of flood damage are described in detail in the interview report (Kumagawa ryuiki jyumin kikitorichosa hokokushu henshuiinkai 2008).

[12] Since sweetfish caught in the Kuma River had been highly valued for their size and taste (Hanaoka 1934; Kosaki 1960), it is regarded as a symbol of the Kuma River.

installed fish ladders at both the Arase and Setoishi dams in 1999, but these did not prove to be adequate countermeasures (Abe 2007).

Furthermore, the dam caused the eutrophication of its reservoir. Inhabitants informed that they were suffering from an offensive smell emanating from the polluted water in the reservoir. Ground vibrations that occurred during discharges from the dam caused cracks in the walls of houses along the dam site.

5.3 Dam Removal Stimulated by Dam Construction Controversies (2000–2007)

The Kawabegawa Dam planned for the Kawabe River, one of the branches of the upper Kuma River, became a nationally prominent controversy in the 1990s.[13] The Kawabegawa Dam was planned for irrigation and flood control; however, it was criticized for being based on an excessive demand prediction by environmental advocates. In spite of those criticisms, both prefectural and national governments and politicians belonging to the Liberal Democratic Party (LDP), the ruling party both at national and prefectural level, strongly promoted these construction projects.

All the cities and towns in the Kuma River watershed had also favored construction, but some residents of the Sakamoto Village who had suffered from the negative impacts of the Arase Dam voiced their opinion that the people in the Kuma River watershed should have the chance to clearly express their attitudes for or against the Kawabegawa Dam. They campaigned for a local referendum ordinance and collected enough signatures to have the proposed ordinance discussed at a village assembly. The assembly members discussed it at an unusually heated meeting and then rejected the proposal by a 7 to 6 vote in 2001.

Furthermore, in 2001, around 20 members belonging to Kuma River Fishery Cooperative Association organized the voluntary association named "Sakamoto Village Fisherman Association" to advocate against Kawabegawa Dam construction.[14]

Coincidentally, the permit for hydropower generation at the Arase Dam granted for 50 years in 1953 had an expiration date of March 2003. Since the hydropower generation at Arase Dam has been run by Kumamoto Prefecture, the prefectural government needs a permit to withdraw water from Kuma River. The permit has been granted by the national government as the River Administrator of Kuma River on condition that its water use does not make adversarial effects on other users. As that date approached, residents of the Sakamoto Village began expressing objections to the permit renewal and to request the removal of the dam instead. They organized the association for advocating the dam removal. Fishermen along the

[13] For more details on the Kawabegawa Dam issues, see Takahashi (2009) or Kumamoto Nichinichi Shimbun shuzaihan (2010).

[14] *Yomiuri Shimbun*, June 16, 2001.

Yatsushiro Sea, including the river mouth of the Kuma River, also requested the prefectural government to remove the Arase Dam.[15]

At first, the Kumamoto prefectural government wanted to renew the license and keep the dam in existence. The prefectural government held a meeting at the Sakamoto Village to explain the position of the prefecture in favor of renewal; at this meeting, a great number of participants objected to the dam and to the permit renewal.[16]

In September 2002, the village assembly adopted a statement requesting removal of the Arase Dam. The statement is sent to the national government and the prefectural government, insisting that Arase Dam caused (1) flood damage, (2) water quality degradation, (3) accumulation of sand and soil, (4) vibration nuisance, (5) loss of downstream river flow, and (6) loss of fishing grounds. The statement clearly described "we strongly request that water license renewal and continuation of Arase Dam should be abandoned." The statement was adopted by unanimous vote at the village assembly.

Those movements at Sakamoto Village influenced the prefectural governments' attitude. Governor Yoshiko Shiotani told that the accepted statement should be in serious consideration and the prefectural government should not stick to the renewal of water license and be flexible.[17] The Sakamoto Village mayor and assembly members visited the governor and directly told that consensus of the village was against the existence of Arase Dam.[18]

The Kumamoto prefectural chapter of the Liberal Democratic Party (KLDP), which held a majority of seats in the assembly, also considered the possibility of the removal in a task force and submitted an opinion brief requesting removal to Kumamoto Prefectural Governor Shiotani. In December 2002, Governor Shiotani announced her decision to remove the Arase Dam at the prefectural assembly, noting that the cost of removal (approximately 4700 million yen) would be less than that of the renewal and maintenance of the existing dam and its associated hydropower generation facility (approximately six billion yen). Governor Shiotani also decided to renew the water license limited only in 7 years and continue the hydropower generation to cover the removal cost in part.[19]

The position of Arase Dam for the Kumamoto Prefecture and Sakamoto Village had largely changed from what it was at 1950s when dam was built. The prefectural share of electricity generated at Arase dam became 0.7%, whereas it was 16% at the initial period where Kumamoto is suffering from electricity shortage.[20] The Sakamoto Village's tax revenue from Arase Dam and the related subsidy from national government was just 0.6% (approximately 24 million yen) of the total annual budget.[21]

---

[15] *Kumamoto Nichinichi Shimbun*, July 16, 2002.

[16] *Asahi Shimbun*, August 11, 2002.

[17] *Kumamoto Nichinichi Shimbun*, September 27, 2002.

[18] *Kumamoto Nichinichi Shimbun*, October 18, 2002.

[19] Kumamoto prefectural assembly meeting minutes, December 10, 2002.

[20] *Kumamoto Nichinichi Shimbun*, October 30, 2002.

[21] *Kumamoto Nichinichi Shimbun*, November 4, 2002. Unfortunately, the amount of related tax revenue and subsidy remains unclear due to the limited available documents.

5.4 *Change of Local Government Policy and Citizen Protests Against It (2008–2010)*

After Governor Shiotani's decision to remove the Arase Dam, a committee of experts discussed specific procedures for the dismantling and decided on a method and schedule for the project. The process seemed to progress smoothly but then underwent a drastic change along with a change of governor.

In April 2008, Ikuo Kabashima was elected as Kumamoto prefectural governor. Although both Governor Kabashima and his predecessor were supported by the LDP and the Arase Dam's removal was not a topic of debate during the election, the new governor suddenly decided in June 2008 to stop the Arase Dam removal project. The primary reason for his decision was the cost of removal, which had increased to 5400 million yen from the original estimate of 4700 million yen. Since Governor Kabashima made this decision without consulting stakeholders even within the prefectural government, his action caused considerable confusion. Governor Kabashima later said that another reason for halting the removal process was that he had received a request to retain the dam from the Future Energy Collegium, an association of former bureaucrats from the Ministry of International Trade and Industry that had formed to promote hydropower generation during the postwar period.[22]

Residents of the Sakamoto Village and members of the Kuma River Fishermen's Cooperative strongly opposed the governor's decision of not removing the Arase Dam and initiated a campaign against it. They directly lobbied the governor and bureaucrats in charge of the Arase Dam operation several times.

These advocacy activities softened Governor Kabashima's attitude toward the dam's removal. Initially, shortly after announcing his decision to withdraw the previous governor's removal request, he said, "Please abandon the idea that we should complete what we had previously decided. We should reconsider when the overall situation has changed.[23]" However, he subsequently indicated in July 2008 that removal was still a possibility, stating "We will deal with the issue flexibly, and the possibility of the conclusion that we will remove the dam after all is not zero.[24]"

On the contrary, downstream farmers at Yatsushiro City started to insist on the maintenance of Arase Dam. Despite the fact that irrigation water is not from Arase Dam but from Yohai Weir located downstream from Arase Dam, farmers worried about the loss of upstream reservoir. The Land Improvement District organized by farmers submitted the statement for maintaining Arase Dam to the city assembly, and it was adopted in September 2008.

Governor Kabashima appointed a project team in the prefectural government to examine the decision about the removal of the Arase Dam. After a month of examination, they reported that the cost of removal would be much more than the cost of maintenance. According to their estimation, the removal cost is 9100 million yen or more than double compared with the original estimated removal cost at the time of Governor Shiotani's administration.

---

22 *Kumamoto Nichinichi Shimbun*, July 3, 2008; May 5, 2009; March 7, 2010.

23 *Asahi Shimbun*, June 7, 2008.

24 *Asahi Shimbun*, July 1, 2008.

In November 2008, the Governor again reaffirmed keeping the dam in place, observing that, "we would need to inject a vast amount of public funds from the prefectural general account into the removal project. The continued existence of the Arase Dam is the best choice so as not to leave Kumamoto Prefecture in need of fiscal reconstruction." The Governor also proposed the conditions for the dam removal as follows: (1) securing of financing for the dam removal, (2) safety of revetment and roads around the reservoir, (3) alternative solution after removing the dam site that also works as the bridge, and (4) established technology for the removal. At this point, the focus was firmly on the burden imposed by removal costs.

Yatsushiro City Mayoral Election

While the policy for maintaining Arase Dam was reconfirmed in late 2008, opponents continued to lobby for the dam's removal. Two elections in 2009 changed the situation again.

In August 2009, Kazutoshi Fukushima, running on a promise to remove the Arase Dam, won the Yatsushiro City mayoral election. Following this, Yatsushiro City started to actively lobby for the removal. Some members of the Yatsushiro City assembly organized a voluntary confederation for the dam removal in November 2009.[25] The confederation expanded its number of members to about 60, including Diet members and prefectural assembly members.[26]

Opponents' protest movements for the dam removal also became active during this period. The kinds of actors lobbying the prefectural government became even more diverse. Citizen groups mainly comprising Sakamoto Village residents lobbied not only the prefectural government but also both prefectural and national political parties and the Ministry of Land, Infrastructure, Transport and Tourism (MLIT), which is in charge of water-related permits.

Request for the National Government's Financial Support with Removal Costs and Its Refusal

In 2009, the Japanese general election resulted in a change of government from the LDP to a new coalition government including the Democratic Party of Japan (DPJ) and the Social Democratic Party (SDP). Since several principal members of both the DPJ and SDP had visited the Arase Dam before the general election and indicated that they would request financial support for the removal from the national government, the DPJ's Kumamoto prefectural chapter submitted an opinion brief seeking a subsidy in the amount of half of the removal cost to Prime Minister Yukio Hatoyama, a DPJ party leader.[27] Governor Kabashima also raised his expectation for the national government's financial support to remove the dam. He visited Seiji Maehara, Minister of the MLIT, to request the financial support in October 2009.[28] However, Minister Maehara expressed reluctance to provide such a subsidy.

[25] *Kumamoto Nichinichi Shinbun*, November 25, 2009.

[26] *Kumamoto Nichinichi Shinbun*, December 15, 2009.

[27] *Asahi Shinbun*, September 14, 2009.

[28] *Kumamoto Nichinichi Shinbun*, October 15, 2009.

As the prefectural government was seeking financial support from the national government, the permit for the Arase Dam was approaching its new expiration date—March 2010. The permit had been extended for 7 years in 2003 on the presumption that the dam would be removed during that time, pursuant to Governor Shiotani's decision. In January 2010, Minister Maehara expressed his view that the existing license would expire in March end and the new license would need more than 6 months to investigate. He also directly refused a request for financial and technical support on the national level for the removal during a conversation with Governor Kabashima.[29] Then, Governor Kabashima faced a challenging situation that he could not have either the financial support or the possibility to renew the existing water license.

In February 2010, Governor Kabashima expressed his policy that the prefectural government would continue hydropower generation with a new water license in 2 years to earn the removal cost as much as possible. The Kumamoto prefectural government, trying to find a way to keep the permit from expiring, sought to apply to MLIT for a new license without approval by the Kuma River Fishermen's Cooperative. However, the MLIT frowned on the prefecture's application for renewal and indicated that "if the continued existence of the Arase Dam is a prerequisite, the overall procedure would take more than 5 months.[30]

The prefectural assembly also made it difficult for the Kumamoto prefectural government to continue hydropower generation at the Arase Dam. The KLDP, the majoritarian political party in the assembly, proposed the withdrawal of new water license application to prevent the confusion. In March 2010, the assembly decided to delete the next fiscal year's budget for continuing the hydropower generation at Arase Dam.

Facing this barrier to continued dam operations, Governor Kabashima finally announced that the prefectural government would start to remove it in 2012. The dam's gates were gradually opened after the permit expiration date of April 1, 2010. All the gates were fully opened by April 11. This time, the decision was final, and the dismantling of the dam has been in progress since September 2012.

5.5 *Dam Removal and Signs of Watershed Restoration (2010–Present)*

As depicted in Fig. 3.4, the river started to recover its connectivity between upstream and downstream that was formerly divided for almost 60 years by the Arase Dam. Despite the fact that removal work was still in progress, we could confirm various signs of environmental restorations. We can see the river flow running at the former dam reservoir. Environmental monitoring reveals the restoration of water quality

[29] *Asahi Shinbun*, January 15, 2010.

[30] *Asahi Shinbun*, February 3, 2010.

and biodiversity, including the diversity of benthos and fish species. Young and Ishiga (2014) report the environmental improvement of bottom sediment at the downstream tidal flat.

According to our interviews thus far, several inhabitants answered that the "water became clean" after the removal work had started. Some people told that "the smell of reservoir's water was bad[31]" and "fish catch nets soon became dirty with algae[32]"; however, those situations improved after the removal work.

Some types of sweetfish fishing have started to be revived. One is what is locally called "Gakkuri Gake" fishing, where the sweetfish are caught while approaching shallow water for spawning. This fishing started to be revived in 2015 at the past reservoir where used to be under the dammed water. Inhabitants with fishing licenses can do this type of fishing. Before Gakkuri Gake fishing started in October, people moved the relatively big stones away to prepare the appropriate spawning bed for the sweetfish. Another method is sweetfish fishing by decoy. We can see the people enjoying this type of fishing downstream of the former dam site, where the flow was much less than at present.

Fishermen along the river mouth told of the positive effect on seaweed cultivation and shrimp fishing by experiencing the change of water quality into "lively" state.[33] While water quality is considered to be improved, some fishermen told of various detritus and flotsam such as leaves, branches, and driftwood began to flow directly into the sea. They suggested the influence from the upstream forest and its devastation.[34]

Movements for the village revitalization begin to start for the post-dam removal period at Sakamoto Village. NPO SSP (Sakamoto Saisei Project) dealing with issues related to decreasing population at the Sakamoto Village started to offer river boating leisure activities on the restored river. A local company, named "Reborn", started rafting tour business so that people can feel the restored river and its business can contribute to the local economy. Sakamoto Jyumin Jichi Kyogikai, the inhabitants' association for village development, has opened a small restaurant that people can enjoy sweetfish dishes along the restored river and is preparing to set a fishing weir for encouraging tourism. A fishing weir is often used for leisure activities or tourism in Japan. The objective of their association is also to increase the number of visitors and develop the area by utilizing the restored river.

6 Discussion

Looking back the process leading to the Arase Dam's removal, we find an intense interaction among stakeholders; however, the overall modes of governance are not collaborative but conflictual. While the residents along the Arase Dam demanded

[31] Interview, February 2, 2015.

[32] Interview, March 5, 2016.

[33] Interview, September 24, 2015.

[34] Interview, September 24, 2015.

Fig. 3.4 Progress of the Arase Dam removal work
Note: All the pictures are taken by the author. Those are taken at almost the same point on the right bank of the river, directing to the upstream

the removal of it, the prefectural government, especially at Kabashima's administration, wanted to maintain the dam. Those adversarial situations have not changed until the final decision to remove the dam was made. The strategy adopted by those who advocated for the dam's removal was resistance as discussed in environmental and resource governance literature in Japan. They protested against the prefectural government's policy and lobbied government agencies and politicians, broadening their protest network.[35] Those findings suggest that the possible modes of interac-

[35]The strategies they adopted were mainly political campaigning. There was no litigation concerning Arase Dam removal.

tion should be broadened in interactive governance studies. As we review the literature in the previous section, much attention has been focused on collaborative mode of interaction. The Arase Dam case, however, indicated that collaborative interactions are difficult to achieve and resistance is an effective strategy for changing a public policy and saving local resident's livelihoods under a certain circumstance. A key to understand such a seemingly countervailing fact lies in the contextual factors underlying each case. As environmental and resource governance literature in Japan (e.g., Mitsumata and Saitoh 2010) argue, some prerequisites should be met for a collaborative relation. Contextual factors, including those prerequisites for the collaboration, behind the issue are critical to understanding desirable modes of interaction in governance.

We can assume that the following contextual factors affect the modes of governance in this case. First, there clearly existed a power imbalance between those who advocated for the dam's removal and those who advocated the status quo. While the prefectural government that owned the dam had economic, human, and knowledge resources, the inhabitants who campaigned for the dam's removal had limited resources. There was an unfillable gap between the prefectural government and the inhabitants who had suffered from Arase Dam and advocated its removal in their power and resources. For those who advocated the removal, it was an encouraging strategy to expand their supporting network for gaining additional resources.

Second, although the power imbalance itself might not prevent a collaborative relationship, different policy beliefs among stakeholders would result in a confrontational relation. In the Arase Dam case, we can trace the reason for why the resistance strategy was adopted back to the critical difference of policy beliefs on Arase Dam. While those who advocated the removal regarded the dam as a source of pollution, those who advocated the status quo, especially Governor Kabashima, regarded the dam as a source of "clean" energy. Rather, he placed much more emphasis on the prefecture's fiscal health issues. As policy studies literatures (e.g., Sabatier 1988) pointed out, policy beliefs deeply embedded in each actor are difficult to change over time and direct their actions for pursuing the policy in accordance with their policy core beliefs. Findings in this study also confirm the stability of policy core beliefs for a relatively long term. Especially, the policy core beliefs of those who advocated Arase Dam removal have been formed through their own experiences that their livelihoods were threatened by the dam. Since they learn from their own experiences, their policy beliefs were robust. Thus, little room remained for the collaborative relationship to emerge between those who advocated the removal of dam and those who advocated its maintenance.

As discussed in previous studies, interactive governance would be effective for dealing with complex social issues. Nevertheless, this study indicates that we have no other choice to resist rather than collaborate when adversarial government intervenes in the dispute over watershed governance. Interactive governance is sometimes referred to as synonym for collaborative governance (Edelenbos and Van Meerkerk 2016); however, we should reexamine the point that interaction does not necessarily mean collaborative relation. Even if resistance to the existing government is a temporal response and just an initial step toward more collaborative relation, we need to put other possible modes of governance than collaboration in the interactive governance literature.

Now, Sakamoto area faces new governance challenges for the area's development. The Sakamoto area is suffering from depopulation and an aging population even though the river is being restoring by the removal works. How people utilize the restored river for area development is a new issue for the area. In this phase, new organizations are emerging for the same purpose of regional development, and their collaborative relation would be a promising option in the future.

7 Conclusion

We have comprehensively described the policy process and interactions among actors in the Arase Dam removal decision, identifying the contextual factors affecting the modes of governance. To better understand the governance of the dam removal, we could conduct additional interviews with stakeholders in the Arase Dam removal process or pursue two lines of comparative studies. One comparative approach would involve other cases of dam removals. Unfortunately, this is the only instance of large-scale dam removal in Japan, but cases from other countries are available for comparative analysis. A second approach would be to draw comparisons with unsuccessful dam removal campaigns. There are several cases in Japan in which dams have remained in place despite local campaigns for their removal. Such comparative studies would help to clarify the significant factors leading to drastic policy change and would make a valuable contribution to further studies of interactive governance.

Acknowledgment Some fieldworks for this study were jointly conducted with the support of Gaku Mitsumata, Daisaku Shimada, and Kazuki Kagohashi. This study is partly supported by JSPS KAKENHI Grant Number JP 16K16236.

Appendix: Chronology of the Arase Dam construction and removal from the 1950s to 2010

Date(s)	Event
Phase 1: Dam as a symbol of "development" (1960s)	
1950s	Electricity shortage at Kumamoto Prefecture
	Kuma River General Development Plan
1955	Completion of the Arase Dam construction work
Phase 2: Dam as a source of "nuisance" (1965–1980s)	
1965	Sever flood damage around the Arase Dam site and its reservoir
	Criticism by inhabitants for the failure of the dams' flood control
	Degradation of the reservoir's water quality

(continued)

Date(s)	Event
Phase3: Dam removal stimulated by dam construction controversies (1990s–2007)	
Late 1990	Nationwide controversies over the Kawabegawa Dam construction
2001	Proposal for local referendum at Sakamoto Village on the Kawabegawa Dam construction and its rejection by village assembly
2002	Formation of “Arase Dam organization” by inhabitants and fishermen
	Several organizations and inhabitants jointly submitted the petition to remove dam and its acceptance by village assembly
	LDP-K proposed the Arase Dam removal
	Governor Shiotani expresses her decision to remove Arase Dam in 7 years
Phase 4: Change of local government attitudes and citizen protests (2008–2010)	
April 2008	Mr. Kabashima became the new governor of Kumamoto Prefecture
June 2008	Governor Kabashima announced his decision to cancel the Arase Dam’s removal
	Fierce opposition movements and lobbying against governor’s decision by inhabitants
November 2008	Project team at the prefectural government reported the removal cost excess the status quo cost
August 2009	Candidate who advocate the dam removal was elected as new Yatsushiro City mayor
	Changes of national government from LDP to DPJ
January 2010	DPJ government expressed its view that water license of Arase Dam cannot be renewed and will expire in the end of March 2010
February 2010	Governor Kabashima applied new water license for maintaining Arase Dam
	LDP-K proposed cuts in the budgets for maintaining Arase Dam and their proposal was accepted in a unanimous at prefectural assembly
March 2010	Governor Kabashima expressed his final decision to remove Arase Dam
Phase 5: Dam removal and signs of watershed restoration (2010–)	
April 2010	The gate of Arase Dam was opened
September 2012	The removal work had started

Source: Compiled by the author based on the data from interviews and collected newspaper articles

References

Abe, R. 2007. Ikanishite Jinkobutu ha seiji to naruka: Arase damu haiki no seijikatei [How Artifacts Become Politicalization?: Political Process of the Arase Dam Removal]. *Shakaigaku kenkyuka kiyo* [Annals of Graduate School of Sociology] 7: 39–55 [in Japanese].

Ansell, C., and A. Gash. 2007. Collaborative Governance in Theory and Practice. *Journal of Public Administration Research and Theory* 18: 543–571.

Beach, Derek, and Rasmus Brun Pedersen. 2013. *Process-Tracing Methods: Foundations and Guidelines*. Ann Arbor: University of Michigan Press.

Crozier, Michel J., Samuel P. Huntington, and Joji Watanuki. 1975. *The Crisis of Democracy*. New York: New York University Press.

Edelenbos, J. 2005. Institutional Implications of Interactive Governance: Insights from Dutch Practice. *Governance* 18: 111–134.

Edelenbos, J., and I. Van Meerkerk. 2016. Vitality in Interactive Governance: Conditions and Challenges. In *The Challenges of Collaboration in Environmental Governance: Barriers and Responses*, ed. R.D. Margerum and C.J. Robinson. Cheltenham: Edward Elgar Publishing.

Edelenbos, J., N.V. Schie, and L. Gerrits. 2010. Organizing Interfaces Between Government Institutions and Interactive Governance. *Policy Sciences* 43: 73–94.

Fisher, F. 2012. Participatory Governance: From Theory to Practice. In *The Oxford Handbook of Governance*, ed. D. Levi-Faur. New York: Oxford university Press.

Hanaoka, I. 1934. *Kumagawa annai* [Kuma River Guide]. Kumamoto: Nagasakijiro shoten shiten. [in Japanese].

Hori, M. 2011. Koteki gabanansuron no totatsuten: Gabanansu kenkyu no kaiko to tenbo wo megutte [Achievement of the Public Governance Studies: Retrospective View and Vision of Governance Studies]. In *Koteki gabanansuno dotaikenkyu: Seifu no sadoyoshiki no henyo,* ed. Niikawa. [Studies on Dynamics of Public Governance: Transformation of modes of government activities] Kyoto: Minerva shobo. [in Japanese].

Inoue, M. 2004. *Komonzu no shiso wo motomete* [A Quest for the Thought on Commons]. Tokyo: Iwanami shoten. [in Japanese].

Jacobs, A.J. 2003. Devolving Authority and Expanding Autonomy in Japanese Prefectures and Municipalities. *Governance* 16 (4): 601–623.

Jentoft, S., and M. Bavinck. 2014. Interactive Governance for Sustainable Fisheries: Dealing with Legal Pluralism. *Current Opinion in Environmental Sustainability* 11: 71–77.

Kooiman, J., ed. 1993. *Modern Governance: New Government-Society Interactions*. London: SAGE Publications.

———. 2003. *Governing as Governance*. London: SAGE Publications.

Kooiman, J., M. Bavinck, S. Jentoft, and R. Pullin, eds. 2005. *Fish for Life: Interactive Governance for Fisheries*. Amsterdam: Amsterdam University Press.

Koontz, Tomas M. 2004. *Collaborative Environmental Management: What Roles for Government?* Washington, DC: Resources for the Future.

Kosaki, K. (ed.) 1960. *Kumagawa gaido* [Kuma River Guide]. Kumamoto: Kumamoto nichinichi shinbun. [in Japanese].

Kumagawa ryuiki jyumin kikitorichosa hokokushu henshuiinkai. 2008. *Damuha suigaiwo hikiokosu: Kumagawa kawabegawano suigai higaishaha kataru* [Dam Causes the Flood Damages: Voices of Victims of the Flood at Kuma River and Kawabe River]. Tokyo: Kadensha. [in Japanese].

Kumamoto nichinichi shimbun shuzaihan. 2010. *Datsudamuno yukue: Kawabegawa damuwo tou* [Fate of a No-Dam Policy: Questioning the Kawabegawa Dam]. Tokyo: Kadokawa gakugei shuppan. [in Japanese].

Lijphart, A. 1999. *Patterns of Democracy: Government Forms and Performance in Thirty-Six Countries*. New Haven: Yale university press.

Lowry, William R. 2003. *Dam Politics: Restoring America's Rivers*. Washington, DC: Georgetown University Press.

———. 2005. Policy Reversal and Changing Politics: State Governments and Dam Removals. *State Politics & Policy Quarterly* 5 (4): 394–419. https://doi.org/10.1177/153244000500500405.

———. 2009. Policy Changes on Canada's Rivers: Different but Not Isolated. *Review of Policy Research* 26 (6): 783–800. https://doi.org/10.1111/j.1541-1338.2009.00416.x.

Mertha, A.C., and W.R. Lowry. 2006. Unbuilt Dams: Seminal Events and Policy Change in China, Australia, and the United States. *Comparative Politics* 39: 1–20.

Mitsumata, G. 2013. Complementary Environmental Resource Policies in the Public, Commons and Private Spheres: An Analysis of External Impacts on the Commons. In *Local Commons and Democratic Environmental Governance*, ed. T. Murota and K. Takeshita. Tokyo: United Nations University Press.

Mitsumata, G., and H. Saitoh. 2010. Kankyoshigenkanrino kyochisenryaku to teikosenryaku ni kansuru ichishiron: Gyosei no kochokutekitaioukaniaru Toyota shi Inabu jyusan zaisanku no jirei kara [A Consideration on the Collaborative Governance Strategy and Resistant Governance Strategy for Natural Resource Management: On the Basis of a Case Study of Inabu Property Ward Forests in Toyota, Aichi, Japan]. *Shodai ronshu* [Journal of University of Hyogo] 61: 151–178. [in Japanese].

Miwa, D., and G. Mitsumata. 2010. Komonzu wo mamori ikasu senryaku nikansuru ichikosatsu: Kinnenno hogakuteki komonzukenkyu no ryukou ni yosete [A Consideration on the Strategy for the Resistance and Utilization of the Commons]. *Shodai ronshu* [Journal of University of Hyogo] 61: 1–32. [in Japanese].

Miyoshi, N. 2007. *Ryuikikanri no hoseisaku: Kenzen na mizujyunkan to sogoteki ryuikikanri no jitsugen ni mukete* [Law and Policy for the Watershed Management: Toward the Realization of Sound Water Cycle and Integrated Watershed Management]. Tokyo: Jigakusha shuppan. [in Japanese].

Obitani, H. 2003. *Damukensetsu wo meguru kankyoundo to chiikisaisei: Tairitsu to kyodo no dainamizumu* [Environmental Movement and Regional Revitalization Over the Dam Built: Dynamism of Conflict and Collaboration]. Kyoto: Showado. [in Japanese].

Ohno, T. 2013. Institutional Design of Watershed Committees in Japan. In *Local Commons and Democratic Environmental Governance*, ed. T. Murota and K. Takeshita. Tokyo: United Nations University Press.

Reed, S. 1986. *Japanese Prefectures and Policymaking*. Pittsburgh: The University of Pittsburgh Press.

Rittel, H.W.J., and M.M. Webber. 1973. Dilemmas in a General Theory of Planning. *Policy Sciences* 4: 155–169.

Sabatier, P. 1988. An Advocacy Coalition Framework of Policy Change and the Role of Policy-Oriented Learning Therein. *Policy Sciences* 21 (2–3): 129–168.

Sabatier, Paul A., Will Focht, Mark Lubell, Zev Trachtenberg, Arnold Vedlitz, and Marty Matlock, eds. 2005. *Swimming Upstream: Collaborative Approaches to Watershed Management, American and Comparative Environmental Policy*. Cambridge, MA: MIT Press.

Sakamoto Village History Editorial Committee. 1990. *Sakamoto sonshi* [Sakamoto Village History]. Kumamoto: Sakamoto sonshi hensan iinkai.

Seki, Y., A. Masano, and K. Kajiwara 2015. *Shakaiteki kyotsushihon toshiteno mizu* [Water as a Social Common Capital]. Tokyo: Kadensha. [in Japanese].

Takahashi, Y. 2009. *Kawabegawa damuha iranai: Takara wo mamoru kokyojigyo he* [We Do not Need the Kawabegawa Dam: For the Public Works to Protect a Tresure]. Tokyo: Iwanami shoten. [in Japanese].

The British Dam Society. 2010. Types of Dams: Gravity Dam. http://britishdams.org/about_dams/gravity.htm. Accsessed 2 Mar 2016.

Torfing, Jacob, B. Guy Peters, Jon Pierre, and Eva Sørensen. 2012. *Interactive Governance: Advancing the Paradigm*. Oxford: Oxford University Press on Demand.

Tsuru, S. 2013. Nihon hatsuno dam tekkyo no genba karano houkoku: Arase dam kono 1nen(3) [Report from the Field of the first dam removal case in Japan: Annual report of Arase dam (3)]" *Shiranuikai/Kumagawa Ryuuikiken Gakkaishi* [Journal of Society for Shiranui Sea and Kuma River Watershed Studies] 7(1): 31–34. [in Japanese].

Young, S.M., and H. Ishiga. 2014. Assessment of Dam Removal from Geochemical Examination of Kuma River Sediment, Kyushu, Japan. *Environmental Monitoring and Assessment* 186 (12): 8267–8289. https://doi.org/10.1007/s10661-014-4002-4.

Tomohiko Ohno is Associate Professor of Environmental Policy at Faculty of Law, Institute of Human and Social Sciences, Kanazawa University, Japan. His research interests are in the field of Japanese environmental policy and governance. He has been conducting research on public participation in the river planning, social capital and watershed governance, and policy change for the watershed restoration.

Chapter 4
Interactive Participation Under a Fragmented Administration System: Watershed Governance in Zhejiang Province, China

Guohan Liu

Abstract This chapter focuses on several types of participatory mechanism currently observed in watershed management in Zhejiang Province. The watershed roundtable mechanism is a new mechanism with aims similar to that used for Taihu Lake Basin initiated by an expert team; however, there are now diverse types of roundtable meetings in the province. In particular, after the provincial water policy called the Five Water Collaborative Governance was issued, new forms of roundtable meetings have been launched. The Our Water Roundtable, which was organized by an environmental nongovernmental organization (NGO) in Hangzhou City, and the Tie Sha River roundtable, organized by an environmental voluntary group founded by the Environmental Protection Bureau and the Communist Youth League of Hangzhou City, focused on the river issues. This chapter discusses how to lead successful interactive participation, which means one interactive form emerged in this region, beyond the consultative authoritarianism in China, and reviews the processes and characteristics of these roundtables.

Keywords Interactive governance · Watershed governance · Environment participation · Fragmented administration system · Roundtable mechanism · Co-governance · Undemocratic regime · Zhejiang Province · China

1 Introduction

Water is an essential resource, not just for human survival but in a range of processes from agriculture to industries. Nevertheless, nearly all modern countries now face problems associated with water. These problems are rooted in the fact that water is a special common-pool resource (CPR). Cycling around the ecological

G. Liu (✉)
Law and Politics School, Zhejiang Sci-Tech University, Hangzhou, China
e-mail: liuguohan@yeah.net

K. Otsuka (ed.), *Interactive Approaches to Water Governance in Asia*,
https://doi.org/10.1007/978-981-13-2399-7_4

environment, water is easily controlled by modern technologies, even though it seems to flow freely. Water is also viewed as both a private good and a pure public or quasi-public good. Effective and efficient provision of adequate clean water that meets a variety of demands has therefore becoming an important challenge.

According to a report issued by the World Health Organization (WHO), 3.575 million people died from water-related causes in 2002, 99% in the developing world (World Health Organization 2008). Developing countries suffer readily from water pollution and water shortages, having less experienced governments able to deal with complex and comprehensive environmental problems. Since 1978, China has been experiencing long-term economic growth, creating severe water shortages and water pollution problems resulting from past growth policies that overemphasized growth of the GDP. The most critical environmental issues faced by China are the shortage of water in the north and water pollution in rivers, lakes, and seas throughout the country (Imura 2007). As populations increase, water quality becomes increasingly dependent on how those living within a watershed care for their river. The watershed environment of most rivers in China has deteriorated over the past few decades due to rapid industrialization and urbanization.

Water is a political issue (World Water Council 2016). The absence of effective public participation impedes governments from finding a suitable means of resolving water issues. The "China Environment Protection Law" explicitly endowed citizens, legal persons, and other organizations with the right to obtain environment information and participate in and supervise environmental protection measures.[1] However, these public participation rights are far from coming into effect at a national level, because of opaque participation procedures and weak support from local communities.

In early 2014, Zhejiang Province launched an important policy boosting governance of water, the "Five Water Collaborative Governance" (*Wu Shui Gong Zhi*).[2] Under this policy, systematic administrative power is intertwined with scattered civil power, forging new and tentative mechanisms for watershed governance. The "Five Water Collaborative Governance" is a mixture of diversified mechanisms. But, in a relatively short time, it can effectively ameliorate water quality. Moreover, some of these mechanisms have been advocated by the central government at the national level.

In a fragmented administration system, the watershed of any river is simultaneously supervised by a number of functionally diverse and geographically separated agencies, with no single bureaucracy accountable for watershed management. Under such a background, interactive participation can help resolve conflicts, allowing a consensus to be reached and facilitating collective action. In Zhejiang Province, interactive participation has been successfully implemented. The river director mechanism (*He Zhang Zhi*), for example, brought together leaders of local

[1] The "China Environment Protection Law" was issued on December 26, 1989 and revised on April 24, 2014.

[2] The slogan "Five Water Collaborative Governance" (*Wu Shui Gong Zhi*) refers to wastewater treatment, floodwater prevention, pond water drainage, water delivery supply, and water conservation.

governments and civic environment protection activists, while the Green Zhejiang, an active environmental protection civil society organization, held a series of successful "Our Water" watershed roundtables aimed at addressing the environment of 11 rivers within the province. Furthermore, the EPVSGL of Hangzhou City (*Hangzhou Zhi Yuan Fu Wu Zong Dui*), a government-related voluntary organization, implemented eight "Tie Sha River" watershed roundtables in the Tie Sha River watershed. Watershed roundtables allow residents to be educated about water quality as well as helping identify reasonable technological solutions, advocating environmental policies, and encouraging local governmental support. Watershed roundtable can be seen as an interactive governance form, the success of which depends on local knowledge and collaboration, especially under an undemocratic regime.

2 Watershed Pollution Problems Following Rapid Industrialization and Urbanization

In 2014, surface water resources in China amounted for 2626.39 billion cubic meters, with 1998.6 cubic meters of water available per capita.[3] Global per capita water resources were estimated at 6123 cubic meters in 2011, more than three times the amount per capita in China. Considering its large population, China is therefore lacking water resources. To make matters worse, the distribution of water resources across districts is far from even. For example, per capita water resources in Zhejiang Province were estimated at 2057.3 cubic meters in 2014, while in Shandong Province it is only 152.1 cubic meters.

Furthermore, a large area of surface water has been polluted by wastewater discharged from factories and urban populations. Of 972 section water quality monitoring points, the ratios of Grade I, Grade II, Grade III, Grade IV, and Grade V water are 2.8, 31.4, 30.3, 21.2, and 5.6%, respectively,[4] suggesting that nearly one-third of the surface water is poor quality. In terms of underground water, ratios of "Perfect," "Good,, "Not Too Bad," "Bad," and "Very Bad" water in 5118 underwater quality monitoring points across the country are 9.4, 25.0, 4.6, 42.5, and 18.8%, respectively.[5] Thus, more than half the underground water is considered "Bad" or "Very Bad."

[3]According to data in the *China Water Resources Bulletin 2014*, released by the Ministry of Water Resources of the People's Republic of China.

[4]Data cited from the *China Environment Bulletin of 2015* released by the Ministry of Environmental Protection of the People's Republic of China. Surface water quality standards in China are evaluated according to contents of 21 different materials, including COD, biological oxygen demand (BOD), ammonia, phosphorus, total nitrogen, petroleum, and phenols. Grade I represents highest-quality water and Grade V the worst quality. It is generally accepted that Grades I and II are high in quality, and Grades IV and V poor in quality. However, some surface water can even be evaluated as "fails to meet Grade V" or "inferior to Grade V."

[5]Data according to the *China Environment Bulletin of 2015* issued by the Ministry of Environmental Protection of the People's Republic of China.

Table 4.1 Water quality of the main rivers in Zhejiang Province

River	Proportion of Grade I, II, and III water
Qiantang Jiang River	74.5%
Cao'e Jiang River	80.0%
Yong Jiang River	50.0%
Jiaojiang River	72.7%
Qujiang River	100%
Feiyun Jiang River	100%
Shao Xi River	100%
Aojiang River	NA, proportion of Grade II water is 25%
Great Canal	0%
Plain river network	16.7%

Source: Zhejiang Province Environment Protection Bureau (2015), *Bulletin of the Environmental Situation of Zhejiang Province, 2014*

Situated in southeast China, Zhejiang Province belongs to the subtropical monsoon climate zone. Average annual rainfall reaches 1600 mm, and the total annual amount of available water resources is 95.5 billion cubic meters. Zhejiang Province is mountainous, 70% of the 101.8 thousand km^2 territory represented by mountains or hills, 23.25% by plains or basins, and 6.5 by rivers or lakes. The surface water system of Zhejiang Province is composed of eight main river systems (Qiantang Jiang River, Cao'e Jiang River, Yong Jiang River, Jiaojiang River, Qujiang River, Feiyun Jiang River, Shao Xi River, Aojiang River), the Great Canal, and the plain river network. Overall, the water quality of the eight main river systems is relatively better than that of the Great Canal and the plain river network (Table 4.1).

Zhejiang Province is one of the most industrialized areas in China, characterized by a large number of active private enterprises. Most of these enterprises are medium- to small-scale, scattered in villages, towns, and suburbs. They include chemical, electroplating, printing and dyeing, tanning, papermaking, and pharmaceutical sectors. Due to a lack of adequate knowledge and financial support for sewage treatment technologies, some of these enterprises discharge industrial wastewater directly into rivers.

The last 10 years have seen the rapid urbanization of Zhejiang Province. In 2014, the urbanization rate reached 64.87% compared to 54.00% in 2004. As a result, in many cities, the increase in urban population spread has outpaced the growth of domestic sewage treatment facilities. In 2013, the per capita consumption expenditure of rural households in Zhejiang Province reached 11760.2RMB Yuan (about 1795.8 $US), ranking it just below Shanghai and Beijing cities. Consumption patterns of the rural residents are very similar to those of urban populations; however, in many towns and villages, the sewage and garbage disposal systems remain under construction. Without these public facilities, rivers are therefore treated as disposal plants. In addition, pesticides, chemical fertilizers, stockbreeding, and merchant shipping are also major causes of water quality deterioration. Because the source of pollution is so varied, no unitary measure has been effectively implemented to resolve the problem.

3 Fragmented Administration of Water Resources

Since water issues concern such a variety of stakeholders, new concepts and techniques are indispensable for effective water governance. The Integrated water resources management (IWRM) (Teisman and Geert 2013), adaptive governance, and interactive governance have all been advocated in the domain of water resources protection (Mark Lubell and Edelenbos 2013; Claudia Pahl-Wostl 2015). All of these concepts and techniques emphasize the role of substantive public participation in rational decision-making. Thus, citizens are seen as an important resource in watershed management (Morton and Brown 2011: p. 6).

However, in China, public participation remains restricted and controlled under the authoritative regime, with regulatory and administrative functions concerning water governance scattered throughout the administrative system. As shown in Table 4.2, at least ten institutions are responsible for different fragments of water governance. For example, within the jurisdiction of a city government, the Environment Protection Bureau is responsible for water quality as well as supervising industrial wastewater discharge. As a result, the Environment Protection Bureau

Table 4.2 Administrative departments of water governance in Zhejiang Province

Departmental category	Department	Main functions
Functional	Environment Protection Bureau	Be responsible for water quality; supervise wastewater discharge from enterprises
	Water Resources Bureau (Shuili Ju)	Be responsible for construction of water utility facilities
	Agriculture, Forestry and Fishery Bureau	Be responsible for irrigation water, use of pesticides and chemical fertilizers, forest protection, and fishery resources protection
Law-enforcing	Urban Management Bureau (Chengguan Ju)	Investigate and treat illegal behaviors that pollute rivers, ponds, lakes, and other water bodies
Coordinative	Five Water Collaborative Governance (*Wu Shui Gong Zhi*) Office	Coordinate, carry forward, and supervise the various government bureaus, enterprises, social organizations, and communities concerning water governance
	Special Committees on Watershed Governance	Comprehensive institution responsible for special rivers, lakes, and dams
Auxiliary	Sanitation Bureau (Huanwei Ju)	Treat diseases caused by polluted water
	Housing and Construction Bureau	Be responsible for planning and constructing sewage conduit networks
	Transportation Bureau	Supervise transportation on rivers, lakes, and other water bodies
	Landscaping Bureau (Yuanlin Lvhua Ju)	Plant and maintain trees, flowers, and grass around rivers, and construct relevant landscaping facilities

Source: Compiled by the author

is equipped with special instruments, knowledge, and a water monitoring system. In contrast, the Water Resources Bureau (Shuili Ju) is charged with planning and constructing water utility facilities, while the Agriculture, Forestry and Fishery Bureau is in charge of irrigation facilities, use of pesticides and chemical fertilizers, forest protection, and fishery resources protection. These three bureaus form the functional departments directly involved in protecting and ameliorating water quality.

The Urban Management Bureau (Chengguan Ju) is charged with investigating and treating illegal behaviors resulting in pollution of rivers, lakes, and ponds. It is a law-enforcing department, aimed at detection and law enforcement; however, they are lacking in proper knowledge about water resources.

The Five Water Collaborative Governance (*Wu Shui Gong Zhi*) office is in charge of coordinating, carrying forward, and supervising the various government bureaus, enterprises, social organizations, and communities concerning water governance. The Special Committees on Watershed Governance form a comprehensive institution responsible for special rivers, lakes, and dams, in coordination with the Five Water Collaborative Governance department; however, neither has expert knowledge or compulsive power.

The Sanitation Bureau (*Huanwei Ju*) is responsible for treating deceases caused by polluted water, while the Housing and Construction Bureau is in charge of planning and constructing sewage conduit networks. The Transportation Bureau is charged with supervising transportation on rivers, lakes, and other water bodies, while the Landscaping Bureau (*Yuanlin Ju*) undertakes the planting of trees, flowers, and grass around rivers, maintaining and constructing relevant landscaping facilities. In terms of water governance, these two bureaus form auxiliary departments, involved, to a large extent, in affairs affecting water quality.

In addition to the fragmentation of administrative power among various departments, water governance is geographically divided into administrative divisions. As a result, the water at geographical borders is prone to pollution. Furthermore, local governments prefer to use unilateral governance forms to deal with water issues since the law claims that all water resources belong to the nation.[6]

4 The River Director Mechanism

4.1 *Acceptance of the River Director Mechanism (He Zhang Zhi)*

The river director mechanism is an effective method aimed at overcoming the fragmented administration of water governance. The river director mechanism originated in Wuxi City, Jiangsu Province, where from May to June, 2007, a large area

[6]The *Water Law of the People's Republic of China* (revised at the August 2002) claims that all of the water resources belong to the nation and the State Council performs the proprietary rights of all the water resources.

of blue-green algae appeared on Taihu Lake, disrupting the water supply to Wuxi City. In August 2007, the government of Wuxi City implemented the *Section Water Quality Control Objectives and Assessment Methods for the Rivers (lakes, reservoirs, marshes) of Wuxi City (trial version)*, which commanded that the main leaders of local governments and the CPC (Communist Party of China) act as river directors of the 64 rivers in Wuxi City. Since these leaders are able to integrate diverse administrative departments, river directors are deemed responsible for resolving water pollution problems across boundaries.

From 2008, the government of Jiangsu Province further implemented a plan whereby cities and counties within the Taihu Lake watershed are required to partake in the river director mechanism. Some cities and counties transformed the river director mechanism into the "dual river director mechanism *(Shuang Chong He Zhang Zhi*)," nominating two river directors per river, one from the provincial government and the other from the city government. Others went on to adopt a "fourfold river director mechanism (*Si Chong He Zhang Zhi*)," with four river directors per river, representing the city government, county government, town government, and village closest to the river, respectively.

The river director mechanism was the conventional response of the traditional administrative system when faced the challenge of water governance. The traditional administrative system can be seen as a "pressure system (*Ya Li Xing Ti Zhi*)," a typical characteristic of which is the ability to transform certain issues into "political tasks," emphasizing their importance among government leaders (Xuedong 2012). However, the river directors all represent officials of local governments or the CPC, regardless of whether the mechanism is single, dual, or fourfold.

4.2 The River Director Mechanism in Zhejiang Province

The river director mechanism in Zhejiang Province mimicked the methods implemented in Wuxi City. Nevertheless, official river directors are unable to obtain comprehensive real-time information unless residents in the river watersheds are involved. In Zhejiang Province, the river director mechanism started in Haining City, with 30 main government leaders appointed as directors of the 30 main rivers in June 2012. In 2013, Hangzhou City then initiated the river director mechanism, adopting four main initiatives. First, the administrative grids were corrected. All rivers within the city are categorized as provincial level, city level, district level, and street/town level or sections. For example, a river that flows through two or more cities is categorized as a provincial-level river, while a river that flows through two or more districts is categorized as a city-level river, and so on (Fig. 4.1). Bulletin boards describing the implementation of river directors were erected near each river. As shown in Fig. 4.2, the bulletin boards emphasized that all residents were free to contact the river directors at any time.

The second initiative was to confirm the roles of civic river directors. Civic river directors (*Min Jian He Zhang*) were selected from residents living near the rivers,

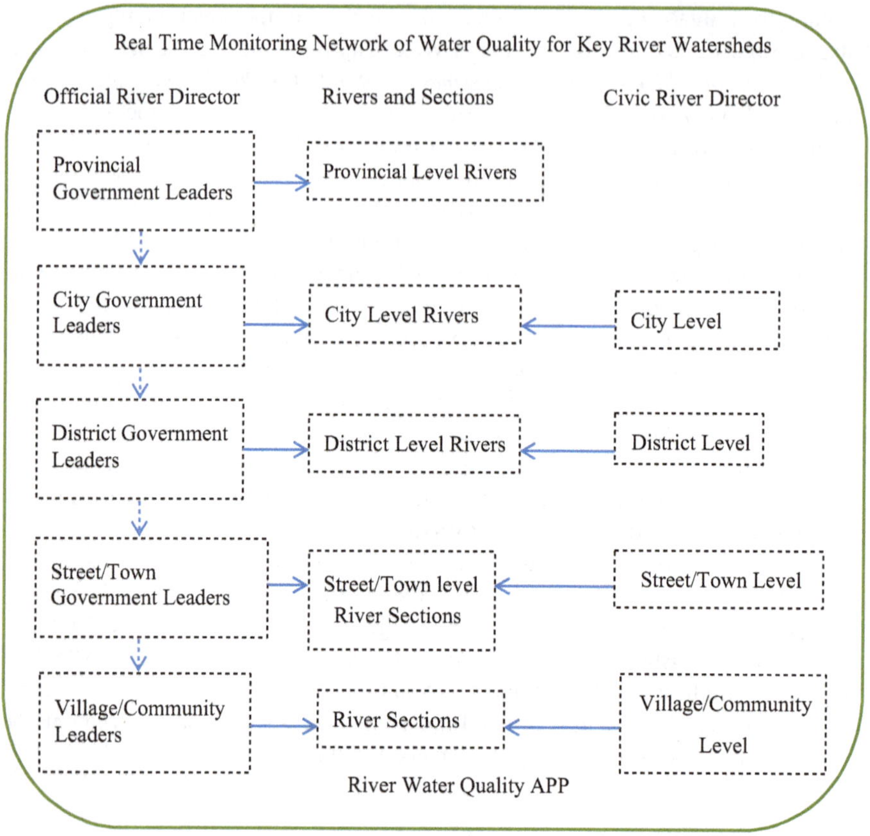

Fig. 4.1 An overview of the river director mechanism in Hangzhou City, Zhejiang Province (Source: Compiled by the author)

environment protection activists, volunteers, and retirees. In April 2014, more than 160 local residents applied for the role of civic river director. Following the interview process, 56 applicants were hired by the city government and charged with supervising 47 rivers within the city as city-level civic river directors. Since then, some district governments (*Qu Zheng Fu*) and street governments (*Jie Dao Ban Shi Chu*) went on to hire additional civic river directors using the same procedures. In the case of special rivers, the civic river director is in constant feedback with the corresponding official river directors.

As a third initiative, further participants were added. For example, the government of Jianggan District forged a "5 + 2" mechanism whereby participants included district-level river directors, street-level river directors, community-level river directors, civic river directors, river policeman, river observers (*He Dao Guan Cha Yuan*), and river cleaners. In addition, some street governments and communities set up grassroots organizations devoted to patrolling and protecting their rivers.

Fig. 4.2 Bulletin board providing information on the river directors (Source: Photo taken by the author (date: 2017, 10th, January))

As a final initiative, IT technologies were improved. New communication technologies can lower participation costs, attracting more residents to participate in water governance. The River Water Quality app is a kind of mobile phone software through which residents can determine the water quality of any river within the city. The River Water Quality app is mainly used by official river directors, civic river directors, water quality observers, river protection volunteers, and ordinary citizens involved in caring for the water quality of rivers. Users can upload photos and directly connect with official river directors using this app.

5 The Watershed Roundtable Mechanism

5.1 Initiation of the Watershed Roundtable Mechanism in China

A watershed roundtable is a group of people with a vested interest in local water quality. Stakeholders, or delegates, express their opinions through the watershed roundtable. Participants then aim to reach an agreement on collective behaviors, although agreement is not compulsory. The watershed roundtable is a potentially effective mechanism aimed at overcoming the fragmentation of authorities concerned with water resources. However, none of the stakeholders have sufficient motivation or the capacity to launch a roundtable, and thus, they are yet to become an exogenous governance form under the current undemocratic regime.

The apathy of stakeholders to participate in a roundtable results from the laws and regulations concerned with water resource management. There are currently three laws concerning the use and protection of water in China: the Water Law of the People's Republic of China (2002 Revision) (the "Water Law"), the Water Pollution Prevention and Control Law of the People's Republic of China (2008 Revision) (the "Water Pollution Law"), and the Environmental Protection Law of the People's Republic of China (2014 Revision) (the "Environmental Protection Law"). None of these laws explicitly confirm the environmental rights of residents. Nevertheless, the Environmental Protection Law does suggest that citizens, legal persons, and other organizations have the right to obtain information, participate in, and supervise environmental protection activities. Moreover, specifically regarding the problem of water, the Water Law stipulates that all water sources are owned by the state, water resources including both surface water and underground water. Thus, according to this law, the water in all rivers is the property of the state. To those devoted to improving the water quality of rivers, it is ironic that they sacrifice their funding and time to protecting an entity that, on paper, does not even belong to them. As a result, many people are unwilling to uncover those who have violated the laws or regulations concerning environment protection.

In 2006, the World Bank initiated a cooperation project with Jiangsu's provincial government in China. This project was entitled, "A Working Outline of the Institution of Environmental Information Round-table Dialogue." Through the support for this project, six community roundtable meetings aimed at water quality improvement in the Taihu Lake Basin were held from 2008 to 2012 (see Chap. 5). Some were even successful; however, the challenge of how to secure the "legitimacy of organizing" community roundtable meetings in the context of the current governance system in China remains (see Chap. 5).

5.2 *The Roundtable Mechanism in Zhejiang Province*

The watershed roundtable mechanism in Zhejiang Province can be traced back to 2009 in Jiaxing City. A series of policies aimed at strengthening the regulation and public participation in environment protection were launched, one of which was the implementation of roundtable meetings. Roundtable meetings were held by relevant departments of the local government before major programs agreed to approve the procedure. Participants included residents, local government departments, journalists, and local employers. The main topics included environment pollution issues, reducing pollutant emission, environment education, and so on (Wei 2015). These roundtable meetings in Jiaxing City were held irregularly by the local government, as a part of government-operated environment policies.

Since implementation by the central government of institutional reform in 2013, certain agencies, nonprofit organizations (NPOs), and nongovernmental organizations (NGOs) began to favor the launch of multi-stakeholder roundtables. For example, in 2010, Hangzhou City TV station initiated a program entitled as "We Roundtable (*Wo Men Yuan Zhuo Hui*)," aimed at encouraging government officers, employers, environmental experts, residents, and journalists to discuss issues of public concern. This program is broadcast weekly every Saturday and Sunday, often addressing environmental and river water pollution problems. However, "We Roundtable" remains a TV program focused on the discussion of problems. The participants are not direct stakeholders of the problems discussed; therefore, no specific resolution or plan is reached.

In early 2014, Zhejiang Province launched an important policy boosting governance of water resources. This policy was termed the "Five Water Collaborative Governance" (*Wu Shui Gong Zhi*) and refers to wastewater treatment, flood water prevention, pond water drainage, water delivery supply, and water conservation. It advocates public participation, encouraging all governments at the city and town level to set up a "Five Water Collaborative Governance Office" (*Wu Shui Gong Zhi Ban Gong Shi*).

Political slogans such as "Beautiful Zhejiang" and "Ecological Construction" used under the Five Water Collaborative Governance were promoted, warranting political correctness of environmental protection through government policies and public opinion. Under these conditions, watershed roundtables were endogenously undertaken by some agencies and organizations. From February to November 2014, the Green Zhejiang, an environment protection civil society organization, organized 11 "Our Water Co-governance" watershed roundtables, each related to the water pollution problems of a specific river within Zhejiang Province. These watershed roundtables were broadcast by Zhejiang Province TV station, increasing public awareness in the short term. However, the Green Zhejiang ended the "Our Water Co-governance" watershed roundtable program in early 2015.

From November 2015, a voluntary organization named the "Environment Protection Voluntary Service General League of Hangzhou City" initiated a series of watershed roundtables concerned with drinking water source protection of Tie Sha River. These watershed roundtables were held at the community level, participants consisting largely of residents of communities near the river. The "Tie Sha River" watershed roundtable was independent of official departments. Although, with time, the operation patterns changed and matured, the "Our Water Co-governance" watershed roundtables and "Tie Sha River" watershed roundtables remain classic examples of public participation in environmental protection in Zhejiang Province (Table 4.3).

Table 4.3 Main watershed roundtables in Zhejiang Province

Name of the roundtable	Period	Main organizers	Main participants	Characteristics
Environment Governance Roundtable of Jiaxing City	From 2009, irregularly	Local government, Environment Protection Association	Local government departments, NGOs, experts	Local government convokes all stakeholders of watershed governance through the roundtable
"We Roundtable" of Hangzhou City	From 2011, weekly	Local government, TV station	Local government departments, experts, residents	TV broadcast
"Our Water Co-governance" roundtable	From 2014 until 2015, monthly	Environmental NGO, TV station	Local government departments, delegates of enterprises, residents, experts	Operated by an environment protection civil organization aimed at resolving specific problems of a specific river
"Tie Sha River" roundtable	From 2015, once every 2 months	Environment Protection Voluntary NPO	Local government departments, resident delegates, community leaders, volunteers	Operated by an environment protection voluntary organization and held in communities surrounding the river

Source: Compiled by the author

5.3 *The "Our Water Co-governance" Roundtable*

5.3.1 The Development and Role of Environmental NGOs

The Green Zhejiang was the first 5A class environmental protection NGO developed in Zhejiang Province.[7] It can be traced back to one of the One Hundred Excellent Volunteer Service Groups rated in June 2000, registered in April 2002 as a chapter of the Green Environmental Protection Society under the Young Volunteers Association of Zhejiang Province. The inability to register as an independent legal entity is a major obstacle of many NGOs in China. The Green Zhejiang was not considered a legal entity for almost 10 years, although core members insisted on conducting environmental protection activities. Finally, in January 2010, the Green Zhejiang successfully registered its first official organization, "Hangzhou City Ecological Culture Society," which was subject to direct administration by the Hangzhou City Environmental Protection Bureau and permitted to organize activi-

[7] According to the current regulations on social organizations, the Social Organizations Registration Agency owns the right to evaluate the performance of social organizations. Based on the evaluation results, social organizations are classified as 1A, 2A, 3A, 4A, or 5A. The 3A classification can be seen as "good" and 5A is considered "perfect."

ties within the city. In 2013, the Hangzhou City Ecological Culture Society, the Hangzhou Low Carbon Science and Technology Museum, the Zhejiang Sunshine Law Firm, as well as the Ruan Junhua, Qi Zhijian, and Xin Hao, among others, jointly established the Zhejiang Green Technology and Culture Promotion Association (Green Zhejiang).

Members are the main supportive sources of an NGO. Although the approximate number of formal members of the Green Zhejiang is only about 400, most of them are enthusiastic activists from various professions and trades. Some are senior civil servants, entrepreneurs, experts, TV celebrities, and academics. In addition, three large-scale alliances also support the Green Zhejiang. The first is the Green Zhejiang Green Footprint Alliance, an organization of enterprises aimed at achieving "low-carbon self-discipline, low-carbon mutual benefits, and low-carbon propaganda." This alliance has attracted a large number of local enterprises with a strong sense of social responsibility and committed to environmental protection. The second is the Green Zhejiang College Students Alliance, which provides an open communication platform for college students to participate in environmental protection, attracting student societies from over 70 universities and colleges. The third is the project team of the Green Zhejiang Roots and Sprouts (Roots and Sprouts), an institution dedicated to the promotion of environmental protection in elementary and middle schools in Zhejiang Province. Roots and Sprouts is an international environmental project aimed at inspiring children of all ages to participate in projects targeting the environment, animals, and communities.

Cooperation with governmental agencies, enterprises, and mass media further adds to the capacity of an NGO. By undertaking governmental outsourcing projects, the Green Zhejiang seeks to build a benign relationship with governmental agencies. Enterprises engaged in environmental protection are willing to cooperate with the Green Zhejiang, and due to strong public concern over environmental problems, mass media is also expected to cooperate with the Green Zhejiang. Through extensive cooperation with such agents, the Green Zhejiang gained the "power" to organize watershed roundtables and promote conflict resolution in a highly complex context.

5.3.2 Organization of a "Our Water" Watershed Roundtable

"Our Water" watershed roundtables have been carried out in 11 river watersheds scattered throughout Zhejiang Province: Beitang River (Hangzhou City), Shan Xia Jin River (Wenling City), Lu San Xiao Xi River (Dongyang City), Long He River (Rui'an City), Fang Men Jiang River (Fenghua City), Xiao Kun Jiang River (Shengzhou City), Shuang Xi River (Anji County), Qi Xi River (Kaihua County), Da Xi River (Lishui City), Guang Chen Tang River (Pinghu City), and Qian Men Fan River (Zhoushan City). The first watershed roundtable pertained to the seriously polluted Beitang River in the downtown area of Hangzhou City in February 2014. It was originally organized as a publicity event, aimed at educating residents on how to dispose of their domestic garbage. Xiacheng District leaders,

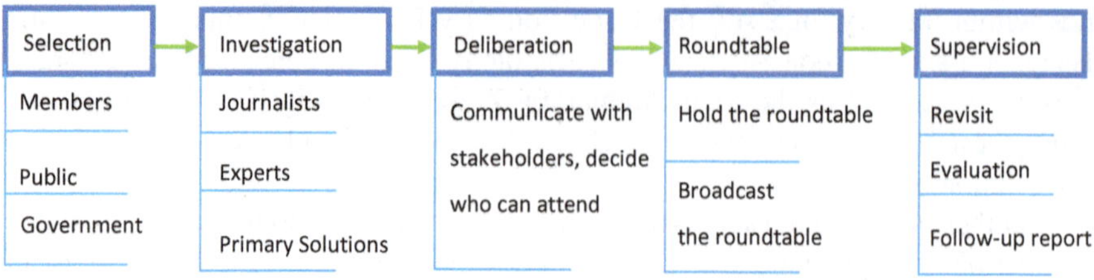

Fig. 4.3 Organization of an "Our Water" roundtable (Source: Compiled by the author)

environmental protection officials, residents living near the river, Zhejiang TV station representatives, and environmental protection volunteers all convened. Unexpectedly, delegates from the Environmental Protection Bureau, the City Construction Bureau, and the Garbage Disposal Company also agreed to take action. As a result, the Green Zhejiang and Zhejiang TV station decided to hold a series of watershed roundtables, following the Bei Tang River model. The organization of a watershed roundtable can be divided into five stages: selection, investigation, deliberation, roundtable, and supervision (Fig. 4.3).

Step 1: Selection

In the first step, the Green Zhejiang selects the river around which the watershed roundtable will be convoked. As an environmental NGO, the Green Zhejiang owns an environmental quality supervision network, consisting mainly of members and volunteers. In addition, residents and the government can provide information on the water quality of rivers in the scope of Zhejiang Province. Since the watershed roundtables are recorded as well as broadcast, the attitude of the local government often becomes the most important standard by which a river will be selected or not.

Step 2: Investigation

In the second step, an investigation group is sent out into the field to determine the actual situation and possible causes of pollution. The investigation group includes environmental volunteers, journalists, experts, and staff from the Green Zhejiang. After investigation, the group can formulate a primary plan of action.

Step 3: Deliberation

In the third step, Green Zhejiang staff deliberate with various stakeholders or their delegates. Since each situation is so specific, there is no homogeneous method by which to ascertain stakeholders. During this process, stakeholders can express their opinions and their preferred solutions.

Step 4: Roundtable

In the fourth step, a roundtable is held near the river. Most attendants have never had the chance to meet so many stakeholders face-to-face. Some are fully aware of the situation and understanding of the opinions of other stakeholders. Most discussions are peaceful; however, many attendants adjust their original standpoints. Nevertheless, most roundtables lead to feasible policies or plans.

Step 5: Supervision
In the fifth step, the effects of the policies and plans are supervised in several ways. Environmental volunteers revisit the rivers periodically and are invited to evaluate the water quality. Furthermore, journalists are allowed to conduct a follow-up report. On January 28, 2015, a large-scale evaluation conference was held in Xiacheng District of Hangzhou City. The attendants included leaders of Zhejiang Province government, leaders of the People's Congress, leaders of the People's Political Consultative Conference, entrepreneurs, voluntary environmental observers, beneficiaries, and local government officials. At this conference, the 11 roundtables were reviewed and the effects evaluated.

5.3.3 Participation and Effects of Watershed Roundtables

To be successful, attendants of a roundtable should feel free to express their true opinions and have a decisive influence. Government officials, journalists, experts, and lawyers were permanent participant of all roundtables. Other participants were invited according to the specific problems of the river. Table 4.4 lists the rivers, sites, dates, main causes, and main participants of the 11 roundtables mentioned above. Here we raise two roundtables as cases.

Long He River Watershed Roundtable
Long He River is located in the Hai'an Community of Tang Xia Town, Rui'an City. In the past decade, Long He River has become seriously polluted by wastewater from small manufacturers and domestic garbage. As a result, the water is discolored and odorous. On May 25, 2014, a watershed roundtable was held by the Green Zhejiang in Tangxia Town. The community leader of the CPC suggested that wastewater from small factories, ineffective governmental regulation, and nearby mine exploitation were the three main reasons for the rivers' degradation. However, the environmental experts pointed out that wastewater from pickling, electroplating, and mold manufacturers was the main cause. The director of the Environmental Bureau of Rui'an City explained that there were 199 pickling factories in the area, only 23 of which passed environmental approval. Delegates of these pickling factories asserted that most were small-scale, consisting of workers who were unable to operate the wastewater disposal equipment, which was also outdated. After the discussion, stakeholders reached a primary agreement whereby the local government formulated a plan to create a centralized wastewater plant. Meanwhile, the enterprises agreed to donate funds to support the Environment Protection Association of Tangxia Town, helping the local government supervise and support factories by updating their wastewater disposal technologies.

Guang Chen Tang River Watershed Roundtable
Guang Chen Tang River flows from Pinghu City to Shanghai City, a length of 16 km. It is severely polluted, having been largely used for shipping, and classified as inferior class V. In recent years, a water hyacinth bloom occurs in spring and summer. Water hyacinth is an aquatic plant with a terrible reproductive capacity.

Table 4.4 Participants and effects of 11 "Our Water" roundtables

No	River	Site	Date	Main causes	Participants
1.	Bei Tang River	Xiacheng District, Hangzhou City	2014/2/21	1.No intercepting sewer	1. Leaders of local governments
				2. Domestic garbage	2. Residents
				3. Wastewater from enterprises	3. Experts
					4. Leaders of the Green Zhejiang
					5. Local TV representatives
					6. Volunteers
2.	Shan Xia Jin River	Wenling City, Taizhou City	2014/3/19	1. Domestic garbage	1. Environment Protection Agency
				2. Industrial wastewater	2. Lawyer
					3. Sewage treatment company
					4. Village chief
3.	Lu San Xiao Xi River	Dongyang City, Jinhua City	2014/4/16	1. Stone processing industry	1. Deputy Mayor
					2. Sewage treatment company
					3. Residents
					4. Manufacturers
					5. Volunteers
4.	Long He River	Tangxia Town, Rui'an City, Wenzhou City	2014/5/27	1. Wastewater from pickling, electroplating, and mold manufacturers	1. Environment Protection Association of Tangxia Town
				2. Domestic garbage	2. Manufacturers
					3. Villagers
					4. Sociologists
5.	Fang Men Jiang River	Fangmen Village, Fenghua City	2014/6/27	1.Swine industry	1. Farmers raising pigs
					2. Villagers
6.	Xiao Kun Jiang River	Shengzhou City, Shaoxing City	2014/7/11	1. Domestic garbage	1.Villagers
				2. Farm product market	2.Peddlers
				3. Fertilizer, agricultural chemicals	3. River cleaners
					4. Sociologist
7.	Shuang Xi River	Anji County, Hangzhou City	2014/7/30	1. Sand and gravel processing manufacturers	1. Farmers of Shuangyi Village
				2. Between two administration jurisdictions of two towns	2. Farmers of Shizhu Village
					3. Owners of sand and gravel processing factories

(continued)

Table 4.4 (continued)

No	River	Site	Date	Main causes	Participants
8.	Qi Xi River	Shuangxi Park, Kaihua County, Quzhou City	2014/9/3	1. Sewage treatment equipment not operated properly	1. Leader of the propaganda department of Kaihua County, CPC
					2. Residents
					3. Alibaba Charity Foundation
					4. Constructor
9.	Da Xi River	Dagangtou Town, Lishui City	2014/11/11	1. Domestic wastewater and garbage	1. River cleaning staff
				2. Coloring matter used by tourists	2. Farmers in the area
					3. Sketching tourists
10.	Guang Chen Tang River	TV station studio, Pinghu City	2014/11/25	1. Excessive growth of water hyacinth	1. Residents
					2. Shipping operators
					3. Channel cleaning companies
					4. River cleaners
11.	Qian Men Fan River	Zhoushan City	2014/11/26	1. Drinking water sources polluted by farm domestic wastewater	1. Residents
					2. Zhoushan Water Affairs Group Co., Ltd
					3. Environmental Protection Companies

Source: Compiled by the author based on field surveys

Water congested with water hyacinths not only kills most aquatic organisms but also damages ship propellers. On November 25, 2014, a watershed roundtable on Guang Chen Tang River was therefore held in Pinghu City to discuss the best way to deal with the water hyacinths. Delegates from the Environment Bureau of Pinghu City claimed that the water was polluted and its nitrogen and phosphorus contents too high. Delegates of the Five Water Collaborative Office further suggested that the fundamental reason for the water hyacinth blooms was rooted in the economic structure of Pinghu City, the pig breeding industry being the main source of income for local residents. Wastewater and feces from pig breeding are directly discharged into the river. Thorough resolution of the water hyacinth problem therefore seemed problematic. As a result, some attendants advised the local government to employ more efficient cleaning companies, while others suggested that the local government formulate a comprehensive plan aimed at improving the water quality of Guang Chen Tang River.

5.4 "Tie Sha River" Roundtable

5.4.1 A Government-Related Voluntary Organization

The Environment Protection Voluntary Service General League (EPVSGL) of Hangzhou City was founded in June 2004 by the Environment Protection Bureau and the Communist Youth League of Hangzhou City. The mission of the EPVSGL is to promote environment protection, defend the legal rights of common persons, and cultivate environment protection consciousness throughout the entire society. In January 2017, the EPVSGL of Hangzhou City consisted of more than 12,000 environmental volunteers, most of whom are divided into 11 branches according to the district/county of their residency. One branch specifically involves participating high school students.

The EPVSGL of Hangzhou City is supported by the Environment Protection Bureau and the Communist Youth League of Hangzhou City, its main office located in the Environment Protection Bureau building in Hangzhou City. A large number of activities are held by the EPVSGL such as environmental promotion at the community level, waste material recycling, environment protection seminars, and so on. The directors of the EPVSGL consist of part-time volunteers; thus, overall the EPVSGL of Hangzhou City can be viewed as a local government-related voluntary organization.

5.4.2 Organization of the "Tie Sha River" Watershed Roundtable

Tie Sha River is located in the Central Zone of Hangzhou City and can be traced back to 861 AD, the Tang Dynasty. It is 6257 m long and varies in width from 25 to 70 m. Since 1931, Tie Sha River became the only drinking water resource in Hangzhou City since the construction of the Qing Tai Men Water Plant. However, with the spread of urbanization, populations near Tie Sha River are increasing rapidly, and as a result, domestic sewage and wastewater from restaurants, laundry, swimming, and fishing have become significant sources of pollution in Tie Sha River. To protect water quality, police monitoring and construction of a protecting fence have been suggested. However, these measures are inefficient and costly and unlikely to prevent the behaviors affecting water quality. One fundamental way to protect Tie Sha River is to enhance environmental protection consciousness among residents near the river. To do so, the directors of the EPVSGL of Hangzhou City initiated a series of watershed roundtables in November 2015 (Fig. 4.4).

Organization of the "Tie Sha River" roundtable consisted of five steps. First was to select a community or a street (*Jie Dao*) near Tie Sha River to act as the main partner. Selection was dependent on the connections between the main directors of the EPVSGL and street (*Jie Dao*) level government or community leaders. The second step was to discuss the theme of the roundtable with the selected partners. Since different streets/communities face different problems associated with pollution of

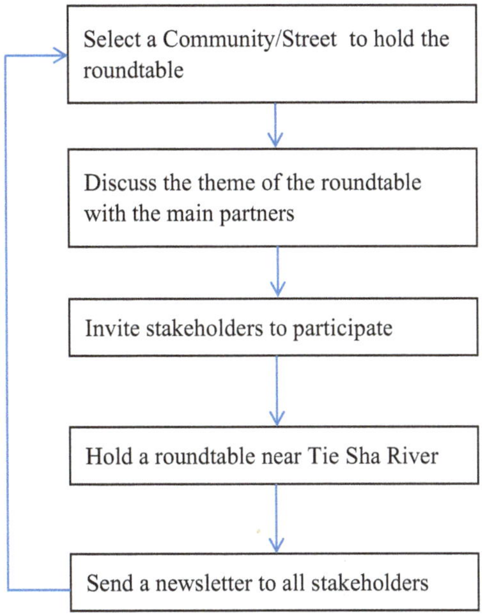

Fig. 4.4 Organization of the "Tie Sha River" roundtable (Source: Compiled by the author)

Tie Sha River, themes were selected accordingly, ranging from water quality, monitoring patrol, laundry management, sewage disposal, and so on. The third step was to invite stakeholders. Some stakeholders take part in only one roundtable, while others are committed to attending all meetings. The forth step was to hold the roundtable, each of which is presided over by one of the directors of the EPVSGL, giving each participant a chance to express an opinion. Most of the time, the directors of local governments are in charge of introducing policies concerning water governance, with volunteers communicating their field experience and findings. In the fifth step, the EPVSGL compiles a newsletter summarizing the roundtable (Fig. 4.5).

5.4.3 Participants and Effects of the "Tie Sha River" Watershed Roundtable

The participants in the "Tie Sha River" roundtables included almost all stakeholders involved in watershed governance of the river. As shown in Table 4.5, participants varied as the scheme changed, with residents near the river, local government officials, and directors of the EPVSGL of Hangzhou City in constant attendance. Moreover, directors of communities near the river, leaders of the voluntary river patrol league (*Zhi Yuan Xun He Dui*), the civic voluntary water quality monitoring center of Hangzhou City (*Hangzhou Shi Min Jian Zhi Yuan Shui Zhi Jian Ce Zhong Xin*), and members of civic observation team (*Min Qing Guan Cha Tuan*) participated frequently. Other stakeholders, such as journalists and water governance experts, took part when relevant.

Fig. 4.5 The eighth Tie Sha river watershed roundtable (Source: Photo taken by the author (date: 2017, 18th, January))

The “Tie Sha River” watershed roundtable is a micro-social movement that takes place at the most fundamental level of the society. It is open to everyone concerned about environment issues related to Tie Sha River. “Tie Sha River” watershed roundtable aims to improve the water quality of the river in a non-confrontational and rather indirect manner. At the first “Tie Sha River” watershed roundtable in Xin Kai Yuan Community, a question pertaining to which bureau was responsible for the water quality of Tie Sha River was raised. This query was subsequently published in the *Hangzhou Daily* newspaper and aroused significant attention surrounding the fragmented administrative system. At the eighth “Tie Sha River” watershed roundtable, the query changed to “who is the river director of Tie Sha River?” Tie Sha River flows through Xiacheng, Shangcheng, and Jianggan districts. Each district- and street-level government in the area has an appointed official river director and civic river directors within their jurisdictions. However, there is no overall river director for Tie Sha River.

“Tie Sha River” watershed roundtable succeeded in raising awareness of the importance of environment protection among residents near the river. As a result, more and more voluntary river patrol leagues were organized at the community level. Students in Dao Mao Xiang primary school also became involved, learning how to monitor the water quality of the river. Moreover, approximately 1 year after the first “Tie Sha River” roundtable, the number of people swimming, fishing, or doing laundry in the river decreased remarkably.

Table 4.5 Primary information on the "Tie Sha River" roundtables

No	Community/street	Date	Main participants
1.	Xin Kai Yuan Community	2015/11/25	1. Leaders of water utilities
			2. Street-level government leaders
			3. Community leaders
			4. Leaders of the EPVSGL of Hangzhou City
			5. Residents
			6. Volunteers
2.	Chao Ming Street	2016/1/27	1. Leaders of water utilities
			2. Street-level government leaders
			3. Community leaders
			4. Readers of the *Hangzhou Daily* newspaper
			5. Leaders of the EPVSGL of Hangzhou City
			6. Residents
			7. Volunteers
3.	Xiao Ying Street	2016/3/30	1. Directors of local government bureaus
			2. Civic river directors
			3. Readers of the *Hangzhou Daily* newspaper
			4. Members of the People's Congress
			5. Members of the People's Consultative Conference
			6. Residents
4.	Jin Lan Chi Community	2016/5/25	1. Directors of local government bureaus
			2.Civic river directors
			3. Engineers of the River Research Institute
			4. Members of the People's Congress
			5. Members of the People's Consultative Conference
			6. Residents
5.	Chao Ming Street	2016/7/27	1. Directors of local government bureaus
			2. Civic river directors
			3. Official river director
			4. Members of the People's Congress
			5. Members of the People's Consultative Conference
			6. Residents
6.	Shang Yang Shi Jie Community	2016/10/12	1. Director of the Water Governance Office of Hangzhou City
			2. Leaders of the Zi Yang Street Government
			3. Community workers
			4. Water rescue team of Hangzhou City
			5. Civic Voluntary Water Quality Monitoring Center of Hangzhou City
			6. Members of the Civic Situations Observer Team
			7. Residents

(continued)

Table 4.5 (continued)

No	Community/street	Date	Main participants
7.	Shang Yang Shi Jie Community	2016/11/16	1. Director of the Water Governance Office of Hangzhou City
			2. Leaders of the Zi Yang Street Government
			3. Community workers
			4. Community Water Protection Team
			5. Members of the City Management Bureau
			6. Residents
8.	Cai He Street	2017/1/18	1. Director the Water Governance Office of Hangzhou City
			2. Director of the Water Quality Monitoring Center of Hangzhou City
			3. Civic river directors
			4. Leaders of the Voluntary River Patrol League
			5. Neighborhood Communist Youth League
			6. Students and teachers of Dao Mao Xiang School
			7. Residents

Source: Compiled by the author based on field surveys

The "Tie Sha River" watershed roundtables were also able to pressurize officials of local governments. During a watershed roundtable, all participants are deemed equal. Local government officials are expected to be open to enquiries from residents and volunteers, a good starting point in resolving the problems related to pollution of Tie Sha River. Positive experiences and methods employed in certain communities are also introduced during roundtables, allowing the leaders of other communities to implement similar strategies.

6 Interactive Participation in Watershed Governance

6.1 Public Participation Under an Undemocratic Regime

Public participation under an undemocratic regime is deemed problematic, since citizens and organizations are not thought to have sufficient power to challenge the government officials who make the final decisions. Appeals system can provide limited political participation for the layperson (Yongsheng 2004). However, the effectiveness of the appeals system depends on the willingness of higher-level authorities to place pressure on abusive or irresponsible local agents, and a term "managed participation" was coined to refer to the participation without liberty or rights (Yongsheng 2004).

Some researchers are much more optimistic about public participation in China, and they consider the current public participation form as "consultative authoritarianism (CA)" that describes the new relationship between the government and civil societies in contemporary China (Teets 2014). They find that local officials have sought to balance the observed benefits of civil society groups under the potential threat of social mobilization, encouraging the formation of autonomous groups while developing a system of positive and negative incentives to guide group activities in desired directions (Teets 2014).

However, the river director mechanism and watershed roundtable in Zhejiang Province show that participation is more than that predicted by Yongsheng (2004) and Teets (2014). At least in the context of watershed governance, the Green Zhejiang and the EPVSGL of Hangzhou City are, to a large extent, free to implement roundtables. Moreover, civic river directors actively participate in water quality monitoring, and therefore, local government officials have no choice but to respond, although civic river directors remain volunteers hired by local governments. The Green Zhejiang and the EPVSGL of Hangzhou City are also free to publicly criticize local government departments despite being closely linked, both personally and officially. Thus, interactive participation between civilians and the bureaucratic system appears to work in a special way.

6.2 *Characteristics of Interactive Participation*

Clean river water is a public good used by all stakeholders living near the river. Rivers often become polluted because of the social dilemmas of stakeholders, each wishing to take action but failing to do so because of opposition. Interactive participation is one possible method of overcoming such dilemmas. During interactive participation, all participants are, to a certain extent, free to express their opinions and give advice. There is no oppressive power placed on the participants. Journalists, experts, residents, volunteers, and environment activists are all free players, thereby offering balance to the systematic power of local governments.

Successful interactive participation can also meet the substantive needs of all stakeholders, acting as an effective tool to offset the shortcomings of a fragmented administrative system. Accordingly, this gives many officials of local government bureaus the incentive to improve performance though interactive participation. Furthermore, interactive participation can help deal with operational problems without conflict. Most stakeholders including the officials of local governments can adjust their behavior or policies at the operational level.

Nevertheless, the results of interactive participation remain uncertain. No tough constraints are placed on the behaviors of stakeholders, and thus, interactive participation remains an informal political process. The results depend on whether the stakeholders can actively reach an agreement. However, this doesn't suggest a powerless system; in fact, by changing residents' behaviors, soft power may be more effective than hard power.

6.3 Factors Leading to Successful Interactive Participation

The river director mechanism and watershed roundtables illustrate successful interactive participation in specific situations. The following four factors are important in creating an interactive participation mechanism.

Factor 1: A Powerful Civil Society Organization
A civil society organization has no influence over local residents or local governments unless it has power. The Green Zhejiang has become the most influential environmental NGO in Zhejiang Province, while the EPVSGL of Hangzhou City has managed to enlist a large number of environmental volunteers. A powerful civil society organization has enough social legality to question government policies or the behaviors of certain enterprises.

Factor 2: Favorable Public Opinion
In recent years, environmental protection has become a consensus among the public. Most local residents, business owners, and government officials affirm the necessity to improve water quality, even though some of their benefits will be impaired. Interactive participation therefore means certain stakeholders having to change their behaviors. Without favorable public opinion, such stakeholders will not feel obligated to do so.

Factor 3: Partners in Mass Media
Mass media can amplify trivial problems into public issues. Throughout all of the "Our Water" roundtables, the TV station of Zhejiang Province remained a partner. By doing so, not only does it provide the studio with a program, but it also shows social responsibility. The TV station of Zhejiang Province is a public TV station, seen as a part of the government of Zhejiang Province. By broadcasting the statements made at roundtables, they can be somehow transformed into promises to the public. In the "Tie Sha River" roundtables, *Hangzhou Daily* newspaper was also an important partner, supporting their activities.

Factor 4: Needs of the Local Community
The most important factor is whether the watershed roundtable meets the needs of the local community. At the end of each watershed roundtable, local governments promise to implement a series of policies and plans. However, the attitudes of the local government also decide whether a watershed roundtable can be held within its jurisdiction. When a watershed roundtable is held, it suggests that the local government is willing to accept this new form of public participation.

7 Conclusions and Discussions

Due to rapid industrialization and urbanization, many rivers in Zhejiang Province have become highly polluted. Domestic garbage, wastewater from small manufactories, feces from pig breeding, and farm chemicals are the main sources. Watershed roundtable meetings can help encourage stakeholders to implement policies aimed at improving water quality. It is widely believed that multi-stakeholder roundtables will fail to become endogenous under the undemocratic regime; however, the "Our Water" watershed roundtables held by the Green Zhejiang and the "Tie Sha River" watershed roundtables held by the EPVSGL of Hangzhou City suggest otherwise that an environmental civil society organization can facilitate the resolution of complex environmental problems by bringing stakeholders together.

The implementation of the "Our Water" watershed roundtables and "Tie Sha River" watershed roundtables cannot be understood in isolation from supportive policies such as "Five Water Collaborative Governance" and the river director mechanism. These two series of roundtables show the emergence of a new form of participation other than "managed participation" and "consultative authoritarianism." These case studies also suggest that interactive participation between civilian and bureaucratic power can occur more frequently and be more effective than assumed until now. Originated as a product of "pressure system," river director mechanism represents a systematic up-to-bottom power. However, more and more local governments adopted measures encouraging coordination between official directors and civilian directors, volunteers, and residents. Watershed roundtables launched by environmental protection NGOs reflect the power of growing civil society. However, these environmental protection NGOs have to seek coordination and mutual understanding with the concerning local governments if they intend to promote the resolution of water pollution problems successfully. For this reason, interactive participation will be consequential resolution for diverse complex environmental problems faced by China, although most of the interactive participation will be implemented with some local characteristics of China.

Support from the TV station of Zhejiang Province was also essential in the success of the "Our Water" watershed roundtables, even though mass media prefers to report new and extraordinary events. When the enthusiasm of mass media fades, civil society organizations do not have sufficient resources to hold roundtables. In fact, the Green Zhejiang has not been able to hold a watershed roundtable since March 2015 and has no plans to do so without the cooperation of the local TV station. Similarly, all of the "Our Water" watershed roundtables were one-off events. Thus, although there are supervisory measures following a roundtable, no permanent institution exists.

Nevertheless, the "Tie Sha River" watershed roundtables were relatively independent and seemed to work in a sustainable manner, helping ameliorate the water quality of Tie Sha River. However, most of the discussion in these watershed roundtables focused on exchanging experiences between participants. Policy problems were not discussed nor were any notable consensus aimed at changing the current situation reached. Thus, if the stakeholders of a watershed roundtable were to be organized into a permanent committee, one with a long-term action plan, they could go on to become a much more powerful interactive governance form, contributing to environmental protection.

References

Cai, Yongsheng. 2004. Managed Participation in China. *Political Science Quarterly* 119: 425–451.

Claudia Pahl-Wostl, Claudia. 2015. *Water Governance in the Face of Global Change: From Understanding to Transformation*. Cham: Springer.

Imura, Hidefumi. 2007. *Environmental Issues in China Today: A View from Japan*. Tokyo: Springer.

Mark Lubell, Mark, and Jurian Edelenbos. 2013. Integrated Water Resources Management: A Comparative Laboratory for Water Governance. *International Journal of Water Governance* 1 (3): 177–195.

Morton, Lois Wright Morton, and Susan S. Brown. 2011. *Pathways for Getting to Better Water Quality: The Citizen Effect*. New York: Springer.

Teets, Jessica C. 2014. *Civil Society Under Authoritarianism: The China Model*. New York: Cambridge University Press.

Teisman, Geert, and Jurian Edelenbos Geert. 2013. Editorial Note on Special Issue on Integrated Water Resources Management. *International Journal of Water Governance* 1: 177–196.

World Health Organization. 2008. *Safer Water, Better Health: Costs, Benefits, and Sustainability of Interventions to Protect and Promote Health*. http://whqlibdoc.who.int/publications/2008/9789241596435_eng.pdf

World Water Council. 2016. *World Water Council 20 Years of Achievement (1996–2016)*. http://www.worldwatercouncil.org/fileadmin/world_water_council/documents/publications/20161128_WWC_book_20years_of_achievement.pdf

Yang, Xuedong. 2012. A Pressure System: A Brief History of a Concept. *Social Science* 2012 (11): 4–12. in Chinese.

Yu, Wei. 2015. *China's Public Participation in Environment Protection*. Beijing: China Environment Publishing House. in Chinese.

Guohan Liu is Associate Professor in the Law and Politics School, Zhejiang Sci-tech University, Hangzhou, China. He is also the Director of the Department of Public Management in the same university. He received his PhD degree in Values and Decision Science in Tokyo Institute of Technology, Japan, in 2009. His research interests focus on local governance modernization, social organization development, and public participation for environment.

Chapter 5
Interactive Governance of Water Environment in Taihu Lake Basin: A Challenge of Legitimacy Under the Authoritarian Regime in China

Kenji Otsuka

Abstract This chapter focuses on the pilot project initiated jointly by Japanese and Chinese research institutes in some communities in an industrial development zone in Yixing City, Jiangsu Province, which is located in the lakefront of Taihu Lake Basin. Here, the large coastal city, Wuxi City, experienced a drinking water crisis due to a huge bloom of blue-green algae in 2007. Subsequently, state, provincial, and local governments took more intensive measures to control water pollution in the basin. However, such top-down governance requires a bottom-up mechanism to be sustainable and effective in the long term. The joint research team conducted eight meetings with the cooperation of the local community leader and grassroots government in their pilot project and identified achievements and difficulties in initiating a bottom-up, interactive mechanism without an official institution. This chapter focuses on the issue of legitimacy in its discussion and introduces the concept of institutional legitimacy to address problems in promoting interactive governance under the authoritarian regime in China.

Keywords Community roundtable meeting · Interactive governance · Top-down governance · Bottom-up mechanism · Legitimacy · Authoritarian regime · Taihu Lake · Yixing City · Jiangsu Province · China

---

K. Otsuka (✉)
Institute of Developing Economies, Japan External Trade Organization (IDE-JETRO), Chiba, Japan
e-mail: kenji_otsuka@ide.go.jp

K. Otsuka (ed.), *Interactive Approaches to Water Governance in Asia*,
https://doi.org/10.1007/978-981-13-2399-7_5

1 Introduction

In China, nearly 40 years have passed since the initial efforts to find solutions to environmental problems were launched. Since the 1990s, various measures have been undertaken, including the enforcement of stricter regulations against sources of industrial pollution, reinforcement of the supervision, and inspection against polluters and local governments through a top-down approach that involved the central government, the People's National Congress, and mass media, and the adoption of a scheme for total emission control of water pollutants in each of the major river-lake basins. Despite all these efforts, violations of environment-related laws and regulations were frequently observed even after entering the twenty-first century. The problem regarding water pollution accidents continues to recur in many parts of the country. Any optimistic view of seeing improvement in the surface water quality of main rivers seemed out of the question (Otsuka 2010a). While the government, in response to the failure of the aforementioned past environment policies, is undertaking a series of countermeasures to enforce stricter rules and expand public investment, mass protests, led mainly by residents suffering from pollution, are arising sporadically in an attempt to directly demand that industries causing pollution as well as the government take immediate actions to solve the environmental problems. Under these circumstances, new initiatives are emerging to establish a new type of water environment governance based on the cooperation among incentivized enterprises and the participation of local residents (Otsuka 2011).

Nonetheless, to examine the possibility of establishing "basin governance" in China based on the participation of diverse stakeholders, there is a need to keep in mind the fact that the political aspects of basins could hamper such an effort (Turner and Otsuka 2005; Otsuka 2008, 2010b, 2012). Each basin in China forms a complex multi-layered political arena. The top layer comprises the state government that administers the basin across local jurisdictions. In the middle are local governments around the basin that claim, respectively, for their own authoritative rights, which are dispersed (split apart). The bottom layer comprises various systems and regulations that impose restrictions on the participation of the residents living along the basin. To build a viable system that conserves and revitalizes the environment around basins in China, a complex set of considerations must be taken into account. The wide range of stakeholders include the government, enterprises, residents, and other interest groups. They have different interests and incentives to realizing an improved water environment, which need to be carefully coordinated in order to reach a consensus regarding how to proceed with the joint efforts to clean the environment. Not only are the stakeholders numerous, but they also represent many social classes. In addition, the relationship between the central and local governments is complex, especially because local governments are very strongly oriented toward growing their regional economies. The country itself is also still lacking in regulatory systems that work democratically. In other words, the institutional building for basin governance in China, which functions as a driver to find possible solutions for water environment recovery, has to be a multifaceted dynamic process that continues to evolve by transforming repeatedly. Simultaneously, the government attempts to reform its administrative systems from the top, voluntary actions emerge from the bottom-seeking self-governance (Ostrom 1990), and negotiations, coordination, and cooperation between communities that work on different vectors come to mutual agreements when interacting with each other.

This chapter will examine the complex basin governance in the Taihu Lake Basin where a water crisis occurred in 2007 followed by new local and state initiatives for water environmental conservation. While examining lake basin governance, this chapter will focus on the top-down mechanism initiated by the government as well as a pilot research project, which was an interactive process, conducted jointly by Japanese and Chinese research institutes from 2008 to 2012. In reviewing the interactive process, this chapter will describe how the pilot project worked and focus on the legitimacy in stakeholders' dialogue as interactive processes under the authoritarian regime in China, referring to discussions about legitimacy in the interactive governance theory (Torfing et al. 2012; Edelenbos and van Meerkerk 2016).

2 Water Environment Governance in Taihu Lake Basin After the Water Crisis in 2007

In mid-April of 2007, blue-green algae bloomed in Lake Taihu earlier than usual. By the end of May 2007, the surface of the water source of Lake Gonghu, where the largest intake to supply water to Wuxi City in Jiangsu Province is located, was literally blackened due to the abnormal massive bloom. By May 29, tap water for drinking and other domestic use had become stale, emitting an abnormal odor and affecting the lives of approximately 2 million citizens to whom this water was regularly supplied. Until June 5, when the municipal authorities declared the city water to be safe, citizens rushed to stores to buy and stockpile bottled water because they could not receive regular water supply from their faucets. As an ad hoc countermeasure, the city selected several sites as tentative clean water supply centers, and the local media continued to report the latest progress in the recovery actions taken by the city government to reduce panic among the citizens (Yang 2008).

From the 1980s onward, the eutrophication of Lake Taihu accelerated as the amount of influent-polluted wastewater discharged from plants, farmlands, and residential areas increased. Due to the massive blue-green algae bloom in the eutrophic lake, local tap water supply faced crises on several occasions in the 1990s (Xie 2008). The water crisis of 2007 became a widely known social issue due to its detailed coverage by both domestic and international media. This incident not only pressured the local and national leaderships to take urgent actions but also caused the water environment policy implemented in Taihu Lake Basin to take a drastic turn.

The water environment policy implemented in the Taihu Lake Basin is experiencing a dynamic process of adjustments to the complex structure of local and national government bodies that continue to affect each other. Meanwhile, local projects and attempts to reform the system allow the mainstream approach to lead and drive the progress of necessary changes (as local initiatives) under the supervision of the state (Otsuka 2010a, b). During the intervening years since the water crisis in 2007, the water environment policy implemented in Taihu Lake Basin has gradually, but definitely, shifted from the policy reform planning stage to the policy reform implementation and coordination stage (Otsuka 2012).

However, when viewed from the standpoint of basin governance, this water environment policy is mainly based on top-down policy reform and project implementation, whereas the bottom-up approach, driven by the disclosure of information and participation of the general public, is limited to sporadic activities. Moreover, the local government is implementing new incentive mechanisms, such as economic incentives and performance assessment systems, to reinforce top-down governance. To ensure the effective functioning of these new mechanisms, there is a need to monitor them extensively and, to do so, the disclosure of information and participation of the general public are essential. Furthermore, past water conservation efforts in Japan suggest that improving the water environment of a lake, which is a closed ecosystem, will inevitably be long term and such a long-term approach will only be successful when the government as well as all the other corporate and private stakeholders communicate and collaborate properly (Otsuka et al. 2011).

In the following sections, we describe how governmental initiatives as top-down governance and pilot community roundtable meetings as bottom-up and interactive governance are functioning in the basin. Following these, we focus on the challenges to integrating these different approaches; that is, the legitimacy of the pilot interactive process under the authoritarian regime in contemporary China. Lastly, we conclude this study by summarizing findings and address some tasks for further study of interactive water governance in China.

3 How Does Top-Down Governance Work?

First, it should be noted that the Comprehensive Water Environment Conservation Plan in the Lake Taihu Basin was stipulated by the National Development and Reform Commission (NDRC) and other related ministries under Prime Minister Wen Jiabao's leadership after the water crisis in 2007. This plan had been prepared as a water pollution control plan in the basin by the State Environmental Protection Agency (reformed as the Ministry of Environmental Protection in 2008) before the crisis. In the Comprehensive Plan, not only point and nonpoint source control measures, including industrial, domestic, and rural waste water control, but also structural adjustment measures, including agricultural, industrial, and urban sectors, are incorporated as necessary components of water environmental conservation in the lake basin. Such measures require coordination across different administrative sectors. To satisfy this requirement, the offices of Taihu Lake Water Pollution Control were set up both in central and local governments.[1] Under the Comprehensive Plan, huge public investment toward those measures was expected.

Under this integration, Wuxi City contributed administrative planning, land use regulations, and construction of a sewage network. With respect to land use regulation, the whole city area of Wuxi is divided into the three grades of the Taihu conservation area. The first grade applies to areas within five kilometers of the lakefront as well as the Taihu Lake surface, where industrial, aquafarming, stock farming, and other activities with a heavy pollution burden to the lake are prohibited or severely restricted. This measure is expected to promote industrial, agricultural, and urban restructuring to secure water environmental conservation in the lakeside city.

Another initiative taken by Wuxi City was the Urban Drainage District Construction Project, which covered the whole city area for the period of 3 years post 2009 after the amendment to the Wuxi City Water Environmental Conservation Bylaw in 2008 after the water crisis in the Taihu Lake. Before the project began, small and distributed point sources of discharged water from household balconies, restaurants, car washes, barber shops, and so on in urban districts of the city were not treated but directly flowed into the lake through rivers and channels. In this project, the city districts were divided into 4172 residential quarters where a drain pipe network was built to gather such small point sources into the main drain pipe of urban sewerage treatment plants. Each project in the quarter was supervised by a local party and government leader. This was a top-down measure to integrate small and distributed point sources for treatment in urban districts of the city.

As a top-down initiative, it should be mentioned that regulation for industrial polluters has been enhanced. Effluent discharge standards of CODcr (chemical oxygen demand by the dichromate method), ammonia nitrogen, total nitrogen, and total phosphorus from the six major industrial pollution sources, including urban sewerage treatment plant, textile and dyeing, paper manufacturing, steel, plating, and food and beverage industries in the basin, have been stricter than ever to be almost comparable with those of developed countries. In addition, standards of levies for those pollutants have been raised or newly added[2] to complement these measures.

Several types of incentive mechanisms for industrial polluters and local governments have been introduced as local government initiatives. We can find five different mechanisms initiated by Wuxi City and Jiangsu Province.

The first is the financial incentive mechanism. Jiangsu Province has set up a special fund to grant and subsidize water environmental conservation projects conducted by city and district governments by pooling 10–20% of an increasing budget (due largely to double-digit annual economic growth each year). In the year the fund

---

[1] To stipulate the Comprehensive Plan, NDRC cooperated with the Ministry of Finance (MoF), the Ministry of National Land and Resources (MNLR), SEPA, the Ministry of Construction (MoC), the Ministry of Transportation (MoT), the Ministry of Water Resources (MWR), the Ministry of Agriculture (MoA), the Ministry of Forestry (MoF), and the State Council's Law Making Office (see p. 2, "preface" in the Comprehensive Plan).

[2] Ammonia nitrogen and total phosphorus in the effluent from urban sewerage treatment plants had not been regulated before.

was set up by the provincial government, it financed about 14 billion yuan; this was equivalent to 17% of 5-year funding for all projects from 2008 to 2012.

The second is "ecological compensation," which was encouraged as a pilot measure after an amendment to the Water Pollution Control Law[3] introduced by the Jiangsu and Zhejiang Provinces. In the part of Jiangsu Province where the Taihu Lake Basin is located, it is required that upper-river governments should pay a certain amount of compensation to down-river governments if the water flow or quality were to be reduced in terms of chemical oxygen demand (COD), ammonia nitrogen, and total phosphorus. The provincial Environmental Protection Bureau (EPB) has to calculate the amount of compensation four times per year and to notify these to city and county governments. The city and county governments have 10 days to pay the notified compensation amount to the provincial financial bureau. In 2009, the total compensation amount in the province was 240 million yuan, but in 2013, it was decreased to 2.6 million yuan or one-tenth of that in 2009. Due to this drastic decrease in the amount of compensation, this incentive mechanism was required to upgrade the standard for its excessive flow while introducing a financial incentive for meeting the standard.

The third is a cap-and-trade mechanism. In the part of Jiangsu Province where the Taihu Lake Basin is located, COD emission trade system (ETS) has been introduced for chemical, dyeing, pulp making, chemical fertilizers, brewing industries, and urban sewerage treatment plants since 2008. Each industry should pay 4500 yuan per ton to the provincial financial bureau to get an effluent permit every year, the amount of which is calculated based on monthly average concentration, actual volume of COD effluent, and so on in the previous year. The following year after the introduction of COD ETS in the Jiangsu part of the Taihu Lake Basin, the total volume of COD from the previous year was reported to be reduced by 35%. The State Council will promote the COD ETS nationwide in 2017; however, it is not clear if this system will work to reduce COD effluents sustainably and efficiently due to poor incentives for industries as well as a skeptical view toward such a trading system under poor enforcement of traditional end-of-pipe regulations.

The fourth is the River Master System (RMS), which was introduced in Wuxi City after the water crisis in 2007 and then spread across the country. Each officer in the provinces, cities, counties, and districts is appointed as the "River Master" (RM) for one river to manage and supervise water environmental conservation projects. Upper levels of the Communist Party and government evaluate the performance of officers at the lower levels in terms of project management and outcomes of their designated rivers. In each city, you can find a plate with the name of the RMs for each river, large or small. This is workable under the authoritarian regime in China; however, it will not be effective unless the public monitors their performance as well as the water environment improvement situation.

The fifth is the so-called "track race" mechanism. Track race means the race played among local governments in China to gain high evaluations from the upper

[3] Water Pollution Control Law was stipulated in 1984, amended in 1996, and amended again in 2008.

levels of the government in terms of the core economic growth indicator, GDP. Luo (2012) points out that "policy objectives such as GDP growth, family planning, social stability and so on are broken down and transferred to lower level governments by top-down mechanism. Whether the lower level government has achieved the goal set by the upper level government is the target of personal evaluation for each officer every year." Using this mechanism, Jiangsu Province has designated some villages as "ecological villages" to evaluate them not by GDP but by the ratio of domestic wastewater treatment population, effluent treatment of large-scale stock farming, area cropped organic farming products, cleaner toilet population, and so on; these indicators are closely related to nonpoint source control in rural areas in the Taihu Lake Basin (Yamada 2012). This mechanism is also feasible under the authoritarian regime in China like the RMS; however, it will be difficult to spread this type of village over the basin unless there is a supportive system that enables all villages to achieve the goal of an ecological village.

These five types of economic and political incentive mechanisms are expected to complement top-down initiatives by the state, provincial, and local governments.

4 How Did the Pilot Community Roundtable Meeting Work?

Since the water crisis in 2007, various policy reforms and comprehensive countermeasures have been put into effect in Taihu Lake Basin by the national and local governments. Following the crisis management and policy reform phases, this region has been in a transitional phase where its activities are becoming increasingly oriented toward new policy implementation and coordination. The current challenges are to figure out how to ensure the viability of policy reforms and comprehensive countermeasures and to maintain their effectiveness. In other words, there is a need to define an appropriate method of managing the environment sustainably for a long period of time. In this respect, one significant attempt to promote dialogue and collaboration among key stakeholders at the grassroots level was the arrangement of "community roundtable meetings."

The term "community roundtable meeting" refers to a mechanism whereby representatives of governments, enterprises, and residents are gathered around one table to discuss and exchange thoughts about local environmental issues. In China, various pilot projects and institutionalized mechanisms to promote the disclosure of information and participation of the general public on environmental policies are being established (Otsuka 2010a, b). Community roundtable meetings originated from a pilot project in 2006 in the Jiangsu Province in cooperation with the World Bank and were later defined more specifically in the "Working Guideline of Institution of Environmental Information Round-table Dialogue" established in 2008 (Wang et al. 2009). Prompted by this pilot project in the Jiangsu Province, the Institute of Developing Economies (IDE) and the Center for Environmental Management and Policy (CEMP), School of Environment, Nanjing University, launched a joint study on the social experiments carried out by the participants of the community roundtable meetings with a common aim to promote dialogue among the governments, enterprises, and residents on matters pertaining to environmental conservation.

One of the places this joint study was conducted in was Yixing City, an industrial county-level city that was faced with the serious problem of polluted substances flowing through the waterways, running through the city, and, eventually, flowing into Lake Taihu. Y District was selected as the site of the community roundtable meeting in this city. The geographical, social, and economic characteristic of the district is that this is an economic development zone that attracted numerous industrial players that have long been in dispute with neighboring farming villages; Y District is also an area that is transforming from a village to an urban community for the purpose of aggregating the blocks of land and using them more efficiently. In this district, new "Shequ" (urban communities) are being built on vacated farmland once inhabited by the farming villagers who have accepted the urban redevelopment plan to move to a newly developed residential apartment for the purpose of making more open lots available for corporate use; these new communities have also emerged as the center of attention for bringing about additional environmental issues attributable to poorly developed infrastructure, in addition to the environmental pollution caused by the enterprises operating in the economic development zone. Upon these characteristics, the research team of Nanjing University had a contact with the leader of the district and got his understanding on the purpose and importance of this pilot project.

From 2008 to 2009, three community roundtable meetings were held in Y District (December 3, 2008, January 8 and August 6, 2009) and one in G Shequ (December 8, 2009). Among them, the meeting scheduled in January 2009 was held mainly to review the matters discussed in the previous meetings in December 2009.

In the years 2010 and 2011, two workshops were held (September 24, 2010 and August 10, 2011) in Nanjing City where members of Chinese and Japanese study groups, including the author (myself), and local keypersons contributing to the building of community roundtable meetings in their localities assembled to find ways to improve the pilot program for organizing community roundtable meetings and to identify and share the issues that needed to be addressed. During these two workshops, the aforesaid participants exchanged opinions on the comprehensive process of how to make the community roundtable meeting work, beginning from how it should be prepared, implemented, and, finally, followed up after the meeting is held. The opinions from the members of the Japanese study group included the introduction of similar experiences of local residents in various parts of Japan. In addition, the China Environmental Forum of the Woodrow Wilson International Center for Scholars, based in Washington DC, has co-organized with CEMP and IDE to hold study tours and workshops at Nanjing (January 2010), Washington DC and Chicago in the USA (August 2010), and Tokyo and Suwa in Japan (December 2011) for the purpose of studying three countries' experiences of public participation in lake basin governance.

Concerning the community roundtable meeting on November 18, 2010, the residents in G Shequ organized a meeting of their own followed by the community roundtable meeting in S Shequ on January 15, 2011. On December 10, 2011, and February 18, 2012, two community roundtable meetings were held in G Shequ. The author participated as an observer in the G Shequ residents' meeting held in October 2010 and conducted post-meeting interview sessions with participants of the three meetings held in November 2010 and in January and December 2011 for the purpose of gaining hands-on information on the latest status of local environment issues (Table 5.1). The major findings through this pilot project are as below.

First of all, among the various pilot programs carried out thus far, community roundtable meetings indicated that the residents and enterprises based in community and the government can enter into a constructive three-way talk. To allow for a dialogue with the enterprises, the fundamental problem was how to coordinate between the government, enterprises, and community leaders to come up with an agreeable method to organize the roundtable meetings, something which will be discussed in greater detail later on. Although the residents in Y District, where many enterprises have gathered, were very keen about discussing the environmental pollution caused by these enterprises, they seem to be cautious about holding a direct dialogue with the enterprises. The residents who participated in the G Shequ roundtable meeting in December 2011 expressed the dissatisfaction they had felt for a long time about the environmental pollution caused by the enterprises; however, they also said that they were reluctant to talk directly to the enterprises because they thought that it was the "role of the village leaders" to hold a dialogue with the enterprises and not theirs. After the roundtable, they came to think that such a direct dialogue was very useful and eager to participate in the next meeting if they would have an opportunity. One of the leaders of Y district also commented that holding a roundtable meeting regularly was a desirable solution to prevent such disputes from escalating and that it was very important for the government, enterprises, and residents to keep communicating among each other. We can say it was not easy to organize the roundtable meeting, but it was considered very useful both for residents and local government to have a direct dialogue with polluters in the meeting if it was set successfully.

Second, the theme in the meeting should be keen issues of residents' concerns to facilitate a meaningful dialogue between residents and other participants. Table 5.1 shows that the theme selected for each community roundtable meeting differed slightly each time. Although these pilot programs have been aimed at involving residents in the process of finding solutions to the water environment issues in Taihu Lake Basin, dialogue with the residents must be considered a priority for any community event to be successful. Another thing that became increasingly clear was that the residents and the leaders of community shared a common interest in investigating the problem of environmental pollution caused by enterprises and poor public management in each Shequ district.

In fact, one of the problems that led to water environment issues in Taihu Lake Basin was the pollution caused by enterprises. Naturally, it was included on the agenda in the first community roundtable meeting held in December 2008. The residents who attended this meeting, however, were more concerned about concrete issues such as the direct adverse effects on the regional environment, rather than on

Table 5.1 Community roundtable meetings in Y district

Date	3 Dec 2008	8 Jan 2009	6 Aug 2009	8 Dec 2009	18 Nov 2010	15 Jan 2011	10 Dec 2011	18 Feb 2012
Objective	CRM	Review meeting	Kickoff meeting / CRM	CRM	Public meeting	CRM	CRM	CRM
Theme	Environmental issues along S river	Review meeting	Water environmental governance in the Tai Lake Basin: Information disclosure and public participation	Environmental issues in G community	Reviewing G community roundtable meetings and remained problems	Agriculture, rural lifestyle, and water environmental conservation in S community	The third G community construction and its environment	Exchanging and sharing of experiences of G community roundtable meetings, dialogue on community environmental problems
Participants	Government 6 (county 2/ district 4) Industries 4 Residents 4 NU 2	Government 6 (county 2/ district 4) Industries 4 Residents 4 Expert 1 NU 2	Government 15 (province 3/city 1/ county 5/district 6) Industries 10 Residents 6 Expert 1 Media 2 NU 2	Government 11 (county 3/ district 8) Industries 4 Residents 12 Media 2 NU 4	Residents 15 NU 1 IDE 1	Government 2 (city 1/district 1) Industries 4 Residents 10 G community 1 NU 3 NGO 1	Government 2 (province 1/ district 1) Residents 21 Other communities 3 NU 3 NGO 1	Industries 3 (service 2) Residents 21 NU 3 NGO 1
	Total 16	Total 17	Total 36	Total 33	Total 17	Total 21	Total 30	Total 28
Chairperson	Chief of environmental division in Y district	Lecturer of NU	Lecturer of NU	Lecturer of NU	Deputy director of G community party	Deputy director of G community party	Deputy director of G community party	Deputy director of G community party

Source: Compiled by the author

the water environmental pollution itself. During the roundtable meetings, residents not only complained about the bad water environment but they also raised angry voices regarding problems like bad odors and soot and dust from industries. Moreover, the theme selected for the S Shequ roundtable meeting held in January 2011 was "Agriculture, Lifestyle in Farming Village and Protection of Water Environment." However, the farmers in this district assembled at this meeting with the common concern regarding the problem of water environmental pollution caused by wastewater discharged from local plants. The organizers of this roundtable meeting claimed that for the conservation of water environment, the treatment of wastewater discharged from plants and households is not the only counteraction that must be considered. Farmers also needed to reexamine and change their conventional farming methods, which relied heavily on the abundant use of fertilizers and agrochemicals. On the other hand, farmers expressed their dissatisfaction with the way wastewater discharged from plants manufacturing chemical fertilizers in Y District was handled, pointing to the damage caused to aquaculture as well as their living environment.

Other issues raised by residents and community leaders at the G Shequ roundtable meeting were related to environmental sanitation, such as draining of wastewater from a remodeled facility that was originally designed as a garage, field burning, and litter scattering in residential areas. Moreover, many of the participants of the G Shequ roundtable meetings held in December 2011 and February 2012 believed that problems related to environmental sanitation were caused mainly by migrant workers who came from outside the city. It should also be noted that there were quite a few remarks stating that the "residents' subjective behavior" might also have been the reason for such sanitation problems arising (Table 5.2). These points were the focus of the second roundtable meeting held in G Shequ in February 2012,

Table 5.2 Perception of environmental sanitary problems in the community

	Community roundtable meeting in Dec 2011		Community roundtable meeting in Feb 2012	
[Major reasons of environmental sanitary problems in the community (plural answers)]				
Government does not take any effective measurements	6	(21)	8	(31)
Industries do not act pro-environmental protection	2	(7)	3	(12)
Local community does not organize residents well	5	(18)	6	(23)
Many problems owe to migrant labors	15	(54)	17	(65)
Local residents act self-aware behavior	12	(43)	10	(38)
Others	0	(0)	0	(0)
No/not efficient answer	4	(14)	0	(0)
Total answers	44		44	

Note: ( ) indicates % among total samples in each meeting
Source: Compiled by the author based on questionnaire surveys

which followed the first meeting in December 2011. Specifically, the problem of remodeling the first floor of the garage occurred as a result of the lack of understanding of the users' needs in the design phase. In the residential area of third-term construction plan undergoing within G Shequ, residents have remodeled their garages into rooms for rural migrant workers. The problem was that the residents were not initially aware about the need to connect the remodeled garage to the drainage pipe. The fact that this recognition has started to spread among the residents can be perceived as one of the accomplishments of holding a roundtable meeting.

Third, it should be noted that this pilot project led to voluntary dialogues by the community. According to the leader of G Shequ, who has cooperated with us on this joint study project in organizing the community roundtable meeting, G Shequ voluntarily ran four separate roundtable meetings of their own prior to the community roundtable meeting organized as a part of the pilot program for this joint project. In these voluntary meetings, held in January, March, April, and November of 2011, residents, government officials, and police officers discussed about various issues in G Shequ, including the environment, sanitation, and social order. The leader of G Shequ said that the residents were hesitant at first about participating in this meeting but became more actively involved as they got used to attending this meeting in subsequent rounds. This case example shows that community roundtable meetings in G Shequ, which began as a pilot program for this joint study, are gradually evolving into an autonomous dialogue mechanism within the Shequ.

The extensive development of community roundtable meetings seen in G Shequ is still one of the few "successful" cases. For a community roundtable meeting to work effectively, several requisites need to be met: there must be one or more serious issues in the community that residents feel a desperate need to have resolved; the leader of the community must have a strong desire to solve these issues; the community must be able to gain a certain level of support from the local government authorities; there must be a local group of experts that can coordinate the participation of all stakeholders in these issues (in the case of this joint study project, the study team of Nanjing University served this role), among others. Nevertheless, even when these requisites are fully met, a large challenge remains: how to secure the "legitimacy" (Sabatier et al. 2005, 280–285; Torfing et al. 2012) of organizing the community roundtable meetings in the context of China's current governance system.

5 The Challenge of Legitimacy in the Pilot Project as Interactive Governance

Legitimacy has been discussed in collaborative and interactive governance as an important aspect to evaluate the effectiveness of such governance compared with representative democracy and hierarchical governing systems. Sabatier et al. (2005,

p. 281) has remarked while synthesizing their case studies on collaborative watershed governance that "legitimacy must be understood as a twofold normative requirement, expressed by both procedural and substantive criteria." They also remind us of the difficulties of evaluating its effectiveness in the short term but the necessity of long-term evaluation, and they say that if it is evaluated in the short term, "stakeholder perceptions of changes in the watershed may be used as a surrogate for data on actual changes in evaluations of substantive legitimacy" (Sabatier et al. 2005, 284).

Torfing et al. (2012, p. 61) argue that interactive governance has "three important sources" of legitimacy: "input legitimacy," "throughput legitimacy," and "output legitimacy." Input legitimacy emphasizes "the composition of the participants," which is relevant to the basic rule of representative democracy in Western developed countries. Throughput legitimacy is derived by "showing the internal policymaking processes," which are expected to represent "commonly accepted normative ideals about fairness, responsiveness, and transparency." Output legitimacy is derived "from providing desirable, useful, and promising policy solutions or from earning a reputation as a successful mechanism of governance or trouble-shooting." In this context, one of the twofold concepts of legitimacy defined by Sabatier et al. (2005), procedural legitimacy, would be equivalent to input and throughput legitimacy, whereas substantive legitimacy would be equivalent to output legitimacy.

Here we will discuss these aspects of legitimacy by looking at the process of community roundtable meetings in more detail. First, *input legitimacy* in terms of participants was not secured throughout the meetings. In this pilot project, we did not emphasize representative democracy but pursued success of dialogue in each meeting. The participants in each meeting were gathered by the local government and community leader upon consultation with the research team in Nanjing University.

Figure 5.1 shows changes in participants in the community roundtable meetings conducted as a pilot project in Y district. Focusing on the constituent participants in each meeting, we have identified six types. The first type seen in the meetings conducted on December 2008 and January 2009 represents the prototype of community roundtable meetings in Jiangsu Province. The meetings included the government, enterprises, and residents and were facilitated by an expert. In this pilot project, members from Nanjing University participated as experts in all meetings. The meeting participants in December 2008 and January 2009 were almost the same because the meeting in January 2009 was aimed at reviewing the first meeting (see also Table 5.1).[4] The second type is seen in the meetings conducted on August 6, 2009, and December 8, 2009. In these meetings, a news reporter was invited and reported on the meetings in the newspaper. However, the news media has not been invited since as it was thought by the local organizer of the meetings that transparency through the media could discourage participants to speak out and, thereby, reduce

[4] At the meeting on January 8, 2009, the former deputy director of Jiangsu provincial EPB attended as an expert besides the members of our research team in Nanjing University.

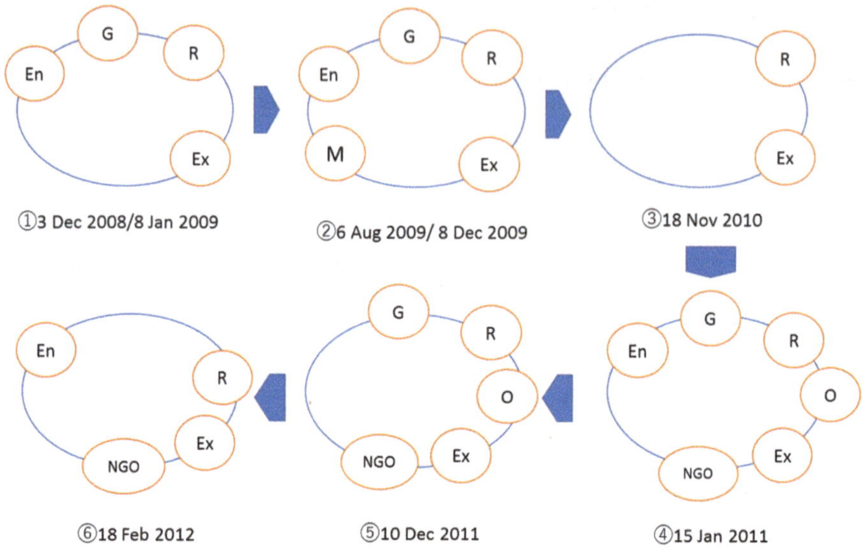

Fig. 5.1 Changes in participants in community roundtable meetings in Y District
G government, *En* enterprise, *M* media, *NGO* NGO, *Ex* expert, *R* residents, *O* other community leader (Source: Compiled by the author)

the effectiveness of the meetings.[5] In the third type seen in the meeting conducted on November 18, 2010, there were only residents and experts from our research team. The aim was to share the experiences of previous meetings among residents, including those who had never participated, and hear about the problems they face in their community. In fourth and fifth type seen in the meetings which were conducted on January 15 and December 10, 2011, respectively, the local organizer invited leaders of other Shequ to share his experiences of community roundtable meetings with them.

The research team in Nanjing University also invited the leader of the environmental NGO (ENGO) based in Nanjing who has built trust through a series of workshops in the USA in August 2010. Since then, the leader or subleader of ENGO has continued to participate in meetings. The ENGO was expected to play the role of assisting the chairperson facilitate the dialogue among participants, especially between residents and other stakeholders based on their experiences in Nanjing and other parts of the province. On the other hand, we find a dearth of participants from enterprises in the fifth type seen in the meeting conducted in December 2011 as well as a lack of participants from the government in the last type seen in the meeting conducted in February 2012. Regarding the reason for the absence of such important stakeholders from the meetings, the local organizer just said that it was not

[5]Local organizers were the deputy director of the Communist Party in the Shequ and staff in Y district. About the trade-off between transparency and seclusion in interactive governance, see Torfing et al. (2012, pp. 208–228).

Table 5.3 Perceptions of results through the community roundtable meetings

Answers	Jan 2011 S		Dec 2011 G		Feb 2012 G	
To obtain environmental information from government and enterprises	10	63%	5	18%	15	58%
To communicate with each other sufficiently under calm circumstances	6	38%	8	29%	12	46%
To express own opinions and suggestions under the fair and neutral chairperson	10	63%	14	50%	5	19%
To solve the problems submitted	3	19%	2	7%	7	27%
To share the common view on the theme	4	25%	0	0%	3	12%
To raise awareness regarding participation and responsibility for environmental protection among the public	–		–	–	6	23%
N/A	0	0%	6	21%	0	0%
Total	33		35		48	

Note: multiple answers available
Source: Compiled by the author based on questionnaire surveys

necessary to invite them because the theme they discussed was not within their jurisdictions. However, it is also true that the local government and enterprises did not always feel positively toward having a dialogue with residents who had a lot of complaints against them.[6] Whatever the reasons, it could be said that the pilot project had failed to maintain input legitimacy throughout the meeting. However, it should be noted that we can find some positive aspects of the involvement of news media and ENGO in some meetings in terms of transparency.

In terms of *throughput and output legitimacy*, we find both positive and negative aspects. In terms of *throughput legitimacy*, it could be said there was a trial-and-error process to achieve meaningful and equal dialogue between residents and other stakeholders including government and enterprises. In the first and third meeting (on December 2008 and August 2009), our research team recognized through onsite observation as well as video records that there were few interactions among them. From the fourth meeting (December 2009), the chairperson had changed the manner of meeting to encourage residents to speak first and then let government and enterprises reply. Since then, the deputy director of the Communist Party in G Shequ followed the chairperson's way of conducting the meeting and active interactions between residents and other participants in following meetings have been observed.

Table 5.3 shows the results of a questionnaire survey for participants in the last three community roundtable meetings that asked their perceptions about results of each meeting. At the January 2011 meeting in S Shequ and the December 2011 meeting in G Shequ, the response "to express own opinions and suggestions under the fair and neutral chairperson" was selected by over half the participants, while the response "to communicate with each other sufficiently under calm circumstances" was selected by nearly half of participants at the February 2012 meeting in

[6] Taken from conversations with local organizers by the author during his onsite visits with them.

G Shequ. These data imply that most participants recognized fair or communicable conditions in the meetings. In the meetings in December 2011 and February 2012, however, an important component of stakeholders did not participate as shown in Fig. 5.1 (the fifth and sixth types, respectively), which could account for the negative responses among participants in the meetings. Especially in the meeting in February 2012, the lack of government participation might have had a negative impact on participants' viewpoints regarding the fairness and neutrality of the chairperson as shown in Table 5.3.

In terms of *output legitimacy*, we have understood some small outputs including the continuous dialogue between residents and the enterprise and the garage sewerage issue found through the meetings as aforementioned in the previous section. Table 5.4 shows some related questionnaire survey results for participants in the meeting compared with those who did not participate in the meeting in the same community. Although more than a half of those who did not participate thought that a community roundtable meeting was a "good way, but not secured of the effect" in environmental protection, over 70 percentage of participants in the meeting thought that the meeting was a "very good way" and might have a "good effect." As shown in Table 5.5, however, positive responses of participants in the three meetings from 2011 to 2012 decreased to less than half. In the January 2011 meeting, the theme was agricultural pollution in the community, but participants among residents had complained about industrial pollution to their cultivated field as well as aquaculture.

As discussed above, we can find both positive and negative aspects in the three sources of legitimacy in community roundtable meetings from the interactive governance perspective. In addition to these types of legitimacy, it should be noted that organizing the meeting itself had been threatened in terms of legitimacy under the authoritarian regime in China. Table 5.6 shows changes in the "legitimacy" of organizing the meetings. Here we can call this "institutional legitimacy," which is a type of legitimacy in terms of degree of institutionalization officially approved under the Communist Party ruling system to secure "social stability" unless the ruling system is eroded. From the launch of the pilot project, the provincial guideline on "the environmental information roundtable dialogue system work" was issued in April 2008 to secure the legality of the pilot project on community roundtable meetings in the Taihu Lake Basin. However, this guideline is just a policy document issued by

Table 5.4 Perceptions regarding the effectiveness of environmental protection by the community roundtable meeting in G community

Answers	CRM Dec 2009		RS Dec 2009	
Very good way, good effect	20	77%	8	30%
Good way, but not secured of the effect	5	19%	14	52%
Merely a formality, no effect	0	0%	5	19%
N/A	1	4%	0	0%
Total	26	100%	27	100%

Source: Compiled by the author based on questionnaire surveys

Table 5.5 Perceptions regarding the effectiveness of environmental protection by the community roundtable meetings (%)

Answers	Dec 2008 CMR	Aug 2009 CMR	Dec 2009 CMR	Jan 2011 CMR	Dec 2011 CMR	Feb 2012 CMR
Very good way, good effect	12(86)	24(77)	20(77)	5(31)	13(46)	12(46)
Good way, but not secured of the effect	2(14)	6(19)	5(19)	11(69)	12(43)	14(54)
Merely a formality, no effect	0(0)	0(0)	0(0)	0(0)	1(4)	0(0)
N/A	0(0)	1(3)	1(4)	0(0)	2(7)	0(0)
Total *n*=	14	31	26	16	28	26

Source: Compiled by the author based on questionnaire surveys

Table 5.6 Institutional legitimacy in the community roundtable meetings in the Y district

Date	3 Dec 2008	8 Jan 2009	6 Aug 2009	8 Dec 2009	18 Nov 2010	15 Jan 2011	10 Dec 2011	18 Feb 2012
Objective	CRM	Review meeting	kick off meeting /CRM	CRM	Public meeting	CRM	CRM	CRM
Chairperson	Chief of environmental division in Y district	Lecturer of NU	Lecturer of NU	Lecturer of NU	Deputy director of G community party	Deputy director of G community party	Deputy director of G community party	Deputy director of G community party
Legitimacy	The provincial guideline*1 →	Environmental democracy*2 →	The state funding project*3 →	Equal dialogue*4 →	→	→	The notice by MoE*5 →	→
Notes	*1: Issued by the provincial EPB in April 2000	*2:The remark by former deputy director of provincial EPB	*3: State Water Special Project at the provincial level	*4: Reforms in the way of meeting, concentrating on residents first	Chairperson has switched to the leader of the community	Since then, NGO participation	*5:Leading statement about nurturing, leading and orderly development of environmental social organizations	The government did not participate

Source: Compiled by the author

the provincial EPB to encourage dialogue to solve environmental issues in the province; it does not have a power to oblige any stakeholder to be involved in the dialogue. Under such a fragile agreement on roundtable meetings in terms of official institution, local governments and enterprises could easily find reasons to avoid burdensome dialogue with residents who complain against them about poor environmental and life quality. As a result, the local organizer of meetings thought it would be better to keep and enhance the institutional legitimacy by adding any remark from those in an official position and organizations. "Environmental democracy" was remarked by a former deputy director of the provincial EPB in the second meeting, while the State Water Special Project funded by the provincial government for the research team in Nanjing University and "the leading statement about nurturing, leading and orderly development of environmental social organizations" issued by the Ministry of Environment (MoE), were also mentioned in the third and the last two meetings, respectively, to remind all participants of the institutional legitimacy of the meetings.

Here it could be raised a question if the institutional legitimacy is a prerequisite for interactive governance with three sources of legitimacy. Interesting to say, we can see some positive aspects of throughput and output legitimacy as mentioned above in the roundtable meetings even under such a fragile condition in terms of institutional legitimacy. It should also be noted that following the remark regarding environmental democracy, the chairperson had stressed the importance of "equal dialogue" among participants at the start of each meeting. This concept, nurtured in the series of meetings, is believed to have kept the institutional legitimacy of these meeting to enable an interactive process with meaningful and active dialogue among participants. Furthermore, it should also be mentioned that the leading statement issued by the MoE was introduced by the Director of the Center for Environmental Propaganda and Education in the provincial EPB. This "hands-on" intervention by a provincial official as a way of metagovernance (Torfing et al., 2012, 134–136) was considered to be effective in securing the institutional legitimacy to enhance interactive process even under the authoritarian regime in China.[7]

6 Conclusion

To achieve environmental restoration of the Taihu Lake Basin that is both adaptive and sustainable in the long run, it must be recognized that top-down governance has limitations in supporting environmental policies enforced in China. In the context of Taihu Lake Basin, changes in the political, economic, and social realms must also be taken into account. From here onward, the main theme to discuss and explore for building new systems should be about how to incorporate a bottom-up mechanism into the new system.

Concerning the bottom-up mechanism, we had conducted a pilot project on community roundtable meetings in one district in Yixing City, near the lakefront. They actually did work in facilitating dialogue between residents and enterprises as well as pushing dialogue between residents and the government on issues about which residents had concerns. However, we can find both positive and negative aspects in the three sources of legitimacy through the meetings. Especially in terms of input legitimacy, it could be said that the pilot project had failed to maintain a balanced representation of stakeholders.

[7] This involvement of the official was based on trust building through a study tour and workshops in the USA in August 2010.

It should be noted that this failure could come from another source of legitimacy—institutional legitimacy—which is a type of legitimacy based on the degree of institutionalization officially approved under the Communist Party ruling system to secure "social stability" unless the ruling system is eroded. The manner of community roundtable meetings is authorized by the tentative guidelines of the EPB and not by the government, so the local organizer had to keep and secure the institutional legitimacy using remarks by a former high-level official on a State-sponsored project and the guideline by the MoE. Interesting to say, however, we can see some positive aspects of throughput and output legitimacy in the meetings even under the fragile condition in terms of institutional legitimacy. It should be also mentioned that "equal dialogue" among participants had been remarked on by the grassroots leader of the community who had been the chairperson in the latter half of the meetings. This remark, nurtured through a series of the meetings, could also have contributed to remind participants of the institutional legitimacy of the meetings to enhance interactive process of dialogue among them.

The pilot project on community roundtable meetings was not led by the government or the community but by the expert team. This could be called "expert-induced interactive governance," referring to government- and citizen-induced interactive governance categorized by Edelenbos and van Meerkerk (2016). This expert-induced interactive governance presents a possibility of changes in attitudes and roles of stakeholders through the power of dialogue under the authoritarian regime in China. It also tells us about the difficulties faced when legitimizing such a dialogue under the fragile formality of meetings. To complement this fragility of institutional legitimacy, we find a kind of self-organized, self-governed attempt in using authorized remarks, state sponsorship, and state ministerial guidelines. The question raised here is how we can get any support from the government to overcome the fragile legitimacy in enhancing any interactive process for collaborative governance. This might depend on how China can reform her regime to be more tolerant of interactive process not only from the top-down way but also from the bottom-up way to nurture the literacy and stewardship of people to govern their own community through dialogue and cooperation.

References

Edelenbos, Jurian, and Ingmar van Meerkerk. 2016. *Critical Reflections on Interactive Governance: Self-organization and Participation in Public Governance*. Cheltenham/Norhampton: Edward Elgar.

Luo, Huanzhen. 2012. Logic of Behavior of Local Government and 'Track Race' in China. *Research on Environmental Disruption* 41 (4): 15–20 (in Japanese).

Ostrom, Elinor. 1990. *Governing the Commons: Evolution of Institutions for Collective Actions*. New York: Cambridge University Press.

Otsuka, Kenji, ed. 2008. *Basin Governance: Issues in China and Japan, Perspectives of International Cooperation*. Chiba: Institute of Developing Economies (in Japanese).

———, 2010a. Tackling Worsening Water Pollution Issues. In *Sustainable Growth: Can China Overcome Her Restraints of Resources and the Environment*, ed. Nobuhiro Horii, 165–195. Chiba: IDE-JRTRO (in Japanese).

———, ed. 2010b. *Governance for Water Environment Conservation in China: Toward Institutional Building in the Tai Lake Basin*. Chiba: IDE-JETRO (in Japanese).

———. 2011. Structure of Governance for Environmental Problem in China, Study Group on China Environmental Problems. In *Handbook of China Environment, 2011–2012*, 50–61. Tokyo: Soso-sha (in Japanese).

———, ed. 2012. *Governance on Water Environment in the Tai Lake Basin, China: Environmental Restoration through Dialogue and Collaboration*. Chiba: IDE-JETRO (in Japanese).

Otsuka, Kenji with Kaori Fujita, Yayoi Isono, and Motoyuki Mizuochi. 2011. A Japanese Experience with Stakeholder Involvement in Water Environment Conservation: The Case of Lake Suwa Basin." Jun Bi, Kenji Otsuka, Junjie Ge and Shi Wang. Stakeholder Involvement in Water Environment Conservation in China and Japan: Building Effective Governance in the Tai Lake Basin. JRP Series No. 155. Chiba: Institute of Developing Economies. http://www.ide.go.jp/English/Publish/Download/Jrp/pdf/155_ch4.pdf

Sabatier, Paul A., Will Focht, Mark Lubell, Zev Trachtenberg, Arnold Vedlitz, and Marty Matlock, eds. 2005. *Swimming Upstream: Collaborative Approaches to Watershed Management*. Cambridge, MA/London: MIT Press.

Torfing, Jacob, B. Guy Peters, Jon Pierre, and Eva Sørensen. 2012. *Interactive Goverance: Advancing the Paradigm*. Oxford: Oxford University Press.

Turner, Jennifer L., and Kenji Otsuka, eds. 2005. *Promoting Sustainable River Basin Governance: Crafting Japan-U.S. Water Partnerships in China*, IDE Spot Survey No. 28. Institute of Developing Economies: Chiba http://www.ide.go.jp/English/Publish/Download/Spot/28.html.

Wang, Shi, Jun Bi, and Junjie Ge. 2009. Practices of Environmental Roundtable Meetings in Jiangsu Province. In *Building Effective Governance for Water Environment Conservation in China: A Social Experiment in Community Roundtable Meetings in the Tai Lake Basin*, JRP Series No. 153, ed. Jun Bi and Kenji Otsuka. Chiba: Institute of Developing Economies http://www.ide.go.jp/English/Publish/Download/Jrp/153.html.

Xie, Ping. 2008. *Historical Development of Cyanobacteria with Bloom Disaster in Lake Taihu: Why Did Water Pollution Incident Occur in Gonghu Waterworks in 2007? After 30 Years, Can We Succeed to Rescue Lake Taihu from Bloom Disaster?* Beijing: Kexue Chubanshe (in Chinese).

Yamada, Nanae. 2012. *Implementation of Rural Nonpoint Source Pollution Control in the Tai Lake Basin: Focusing on the Role of Rural Communities*. In ed. Kenji Otsuka (in Japanese).

Yang, Weize. 2008. *Warning and Action*. Nanjin: Fenghuang Chubanshe (in Chinese).

Kenji Otsuka is Senior Research Fellow in the Environment and Natural Resource Studies Group at the Interdisciplinary Studies Center, Institute of Developing Economies, Japan External Trade Organization (IDE-JETRO), Chiba, Japan. His research interests are in the fields of environmental issues, policies, and social change in China as well as interactive governance on water, environment, and health and disaster risk reduction in Asia. He has been organizing joint research projects with Chinese counterparts, including both governmental research institutes and NGOs since 1993.

Chapter 6
Civil Society and Water Governance in Northern Thailand: Local NGOs and Management of Mekong's Tributaries in Chiang Rai

Siriporn Wajjwalku

Abstract This chapter focuses on two cases of people's organizations in water governance in Northern Thailand and aims to reveal the role of civil society as a stakeholder in water management, including water allocation and flood prevention of the Mekong's tributaries in Chiang Rai. This chapter clarifies the interaction and coordination between local NGOs and government authorities, focusing on participatory opportunities and negotiation capacity. Through the two case studies of the People's Council of Ing River and water allocation and the Association of Chiang Saen Livable City and Kok River Basin Ecology Group and flood prevention by telemetry and early warning systems, this chapter argues the limitations of local NGOs' participation and their negotiation with government authorities in interactive, cooperative way of water governance.

Keywords Water governance · Water resource management · Civil society · Local community · Local NGOs · Decentralization · Participation · International river · Mekong River · Mekong's tributaries

1 Introduction

The Mekong River is the longest river in Southeast Asia (Santasombat 2011). It is 4800 km long and covers 795,000 square kilometers of area, flowing from Southern China to Myanmar, Lao PDR, Cambodia, and Vietnam (Hirsh 2006). Along its length, there are several tributaries in all the riparian countries that have contributed to both development and disaster for people living in the areas. The Kok and Ing Rivers are the two main tributaries of the Mekong in Chiang Rai, Northern Thailand, that have affected the lives and prosperity of people in the past and continue doing

S. Wajjwalku (✉)
Faculty of Political Science, Thammasat University, Bangkok, Thailand
e-mail: wsiripor@tu.ac.th; siriporn.waj@mfu.ac.th

K. Otsuka (ed.), *Interactive Approaches to Water Governance in Asia*,
https://doi.org/10.1007/978-981-13-2399-7_6

so at present. As a precious resource, water from the Mekong and its tributaries is necessary for the daily consumption and economic production of both agriculture and industry; so, a key question is how to manage and allocate this resource fairly to all sectors and stakeholders. In addition, an unexpected and unpredictable amount of water will cause deterioration of the lives of people and the riparian states; another key question, then, is how to prevent flooding and manage the river properly.

Based on the idea proposed by Santasombat that water management needs a cross-scale and inclusive decision-making process at all levels (Santasombat 2011), this chapter extends Santasombat's study, aiming to explore the role of civil society as the stakeholder in the process of water management, including water allocation and flood prevention, for Mekong's tributaries in Chiang Rai. It argues and elaborates on the coordination among local NGOs and government authorities in terms of political power and administrative structure, as well as the interaction between local NGOs and government authorities, focusing on participatory opportunity and negotiation capacity. OECD's, GWP's, and WGF's concepts of water governance will be applied as a framework. Two case studies, namely, the People's Council of Ing River and water allocation and the Association of Chiang Saen Livable City and Kok River Basin Ecology Group and flood prevention by telemetry and early warning systems, have been examined. Field research was conducted, and documentaries were made to serve the purpose of the study.[1] The main argument of the chapter is that without local people's awareness of their rights, as well as unified, strong, and knowledgeable local NGOs, it is very challenging for local people and communities to exercise their rights under the centralized administrative structure that does not allow for water governance, in particular the participation and negotiation among actors involved.

The chapter consists of six parts starting with the introduction that shapes the research question and frames the entire chapter. Following the introduction, the concept of water governance is reviewed briefly in the second part (Sect. 1). In the third part (Sect. 2), the administrative structure of water management in Thailand, particularly in the north, is examined to gauge its nature as an obstacle to water governance in terms of inclusive decision-making and stakeholder engagement. The fourth and fifth parts (Sects. 3 and 4) elaborate the case studies of local NGOs in Chiang Rai that have been active in water management of Ing and Kok Rivers to understand the limits of local NGOs in terms of negotiation capacity and coordination skill in dealing with government authorities regarding water allocation and flood prevention. Lastly, the conclusion section presents the factors that contribute to the ineffective water governance, namely, the centralized administrative structure, nature of the issues, capacity of local NGOs, and awareness of local people and communities. Suggestions for solutions are also put forward.

---

[1]The 3-year project titled "Water and Sustainable Development: Civil Society and Water Management of Mekong Tributaries in Northern Thailand" (2014–2016) supported by the Network for International Development Cooperation (NIDC), The Asia Foundation, and Thailand Research Fund.

2 Water Governance: What Is It About?

In 2000, the World Water Forum in The Hague stated that the water crisis is often a governance crisis and calls for methods and tools to ensure good governance (Rogers and Hall 2003). Recently, good governance has become a key concept and approach for effective water management in many countries and regions including the Mekong and its riparian states. However, as Hirsh (2006) mentioned, the definition and agenda of governance vary depending on different contexts, and stakeholder's interests are also too complex owing to various groups of stakeholders. Therefore, governance requires an integrated approach to managing resources and also implies that inclusive stakeholder engagement is crucial. Hirsh further suggested that the process of and ability to foster negotiation for more sustainable, equitable, and productive use and management of water should be seen as a product of water governance rather than as a "best practice."

Emphasizing the process of participation and negotiation among the actors involved responding to regional and global socioeconomic and environmental changes, Imamura (2007) addressed the need for a more democratic approach to water governance. As governance involves decision-making, which is related to political power and administrative structure, considering rights and justice is inevitable. Badenoch et al. argued, "securing the rights of people with claims over water resources requires governance structures that are inclusive and not only create and allocate rights to different sectors of society, but, more importantly, protect them in the face of competing interests" (Badenoch et al. 2012, p.7). This view is supported by Santasombat, especially regarding the situation in the Mekong region, where the development and management practiced by states with a centralized and top-down policy and implementation without consultation or participation by local people and communities failed. He argued that the local natural resources should be managed by local people or communities for the following reasons. First, the resource itself is local; therefore, it is best managed by local people. Second, as their lives depend on local resources, local people have the requisite knowledge to manage these resources properly. Third, the local control of resources is crucial for a check and balance with the government's centralized administrative system and management. However, given the limited capacity of local entities, Santasombat suggested that "neither purely local level management nor purely higher level management works well by itself. Rather, there is a need to design and support cross-scale management, linking institutions both horizontally – particularly at the local level – and vertically, that is both nationally and internationally" (Santasombat 2011, p.14–15).

Since early 2000s, international organizations and international forums have provided the concepts and approaches of water governance. The Global Water Partnership (GWP) defines water governance as "*the range of political, social, economics, and administrative systems that are in place to develop and manage water resources, and the delivery of water services, at different levels of society*" (Roger and Hall: GWP-TEC 2003, p.16). Rogers and Hall elaborated that the notion of water governance includes the ability to design public policy and an institutional framework that are socially accepted and mobilize social resources to support them.

The process of water policy formulation must aim toward sustainable management of water resources, while the implementation process must be made effective with the involvement of all actors and stakeholders. In addition, as water is a scarce resource, equitable allocation and efficient management will require political drives (Rogers and Hall 2003). This chapter focuses on the political aspect of water policy to understand whether and how different degrees of political power and influence contribute to policy formulation and implementation.

The Water Governance Facility (WGF), which is a collaboration between the UNDP and Stockholm International Water Institute (SIWI) working on knowledge and capacity development related to multiple thematic areas, including integrated water resources management (IWRM), transboundary waters, water supply and sanitation, gender equality, water integrity, and climate change adaptation, shares a similar idea of water governance. For WGF, "water governance refers to the political, social, economic and administrative systems in place that influence water's use and management, essentially who gets what water, when and how, and who has the right to water and related services, and their benefits."[2] It also asserts that "governing water includes the formulation, establishment and implementation of water policies, legislation and institutions, and clarification of the roles and responsibilities of government and civil society, and the private sector in relations to water resources and services."[3] Among the four dimensions of water governance, namely, social, economic, political, and environmental, this chapter pays attention to the political dimension, which emphasizes "equal rights and opportunities to take part in decision-making process."[4] In practice, this dimension emphasizes the participation of common and marginalized people in decision-making, implementation, and conflict resolution. With the assumption that common and marginalized people should have equal rights and opportunities to take part in the decision-making process, this chapter will explore whether and how such people in Chiang Rai are able to exercise their rights in the water management process.

OECD has recently given priority to water governance, in particular, the stakeholder engagement, as an important principle. The OECD Principles on Water Governance provide a framework to understand whether water governance systems are performing optimally and help to adjust them where necessary. Under this principle, three main elements are emphasized, namely, enhancing the effectiveness of water governance, enhancing the efficiency of water governance, and enhancing trust and engagement in water governance. This chapter will look at the last element, "enhancing trust and engagement in water governance," which includes Principle 10. Principle 10 aims to "promote stakeholder engagement for informed and outcome-oriented contributions to water policy design and implementation" (OECD 2015). This chapter will examine how local NGOs interact and negotiate with government authorities in order to propose and have their demands incorporated into the water policy and plan.

---

[2] Water Governance Facility (WGF): watergovernance.org/water-governance/

[3] Water Governance Facility (WGF): watergovernance.org/governance/what-is-water-governance/

[4] Ibid.

3 Water Management in Northern Thailand

According to the Department of Water Resources (DWR), Ministry of Natural Resources and Environment (MoNRE), Thailand, at present, there is no law or act to govern or manage water resource.[5] In 1989, the government issued the Regulation of the Office of the Prime Minister on National Water Resources Management[6] which was revised twice in 2002 and 2007. This Regulation has provided the mechanism and guidelines to direct, administrate, control, and supervise the water resource management policy and plan. With this Regulation, two important committees, namely, the National Water Resources Committee (NWRC) and River Basin Committee (RBC), were established. The DWR functions as the secretary of both committees (DWR 2014).

The NWRC is a national-level organization appointed by the cabinet and chaired by the prime minister. The members of the committee consist of assigned deputy prime ministers, permanent secretaries, director generals, technical experts, etc. The committee mainly oversees the holistic policy and strategic plan of national water resource management including initiating and amendment of related rules and regulations; making policy and plan; budgeting; inter-agency coordinating, monitoring, and evaluation of decent authorities; and supporting RBC in policy implementation at basin level.

Regarding RBC, this committee was established according to the Regulation of the Office of the Prime Minister on National Water Resource Management in 2007. There are 25 RBCs all over the country following the main 25 river basins in Thailand (see Appendix 1). Each committee's members appointed by NWRC consist of representatives from government agencies, private sector, and civil society, with the amount of no more than 35 persons. Each RBC oversees and manages water resource of the basin including making policy and plan, budgeting, coordinating between central government agencies and local authorities, prioritizing development projects and allocating water, monitoring and evaluation, settling conflicts and solving problems, as well as coordinating with other RBCs. Under each RBC, subcommittees, working groups, and networks are set up to function at river and subriver basins as well as the provincial level. Chart 6.1 displays the structure of NWRC and RBC.

Interestingly, while NWRC and RBCs are designed and assigned to manage water resource at both national and basin levels, according to DWR, there are other 12 ministries/agencies with more than 30 departments involved in this issue (see Appendix 2). Each ministry and agency has its regulation to govern its policy and plan, as well as its budget to implement its projects and activities. In addition, due

[5]When this manuscript was written (October 2017), the draft of Water Resource Law was under consideration of the National Legislative Assembly after it was submitted in May 2017.

[6]The term for this regulation varies according to the organization involved. This regulation was issued by the Office of the Prime Minister, and it was called "Regulation of the Office of Prime Minister." However, in DWR's website and document, the term "The Prime Minister Regulation" was used. In Thai, both organizations used the same word, "Rabieb Samnak Nayokratthamontri."

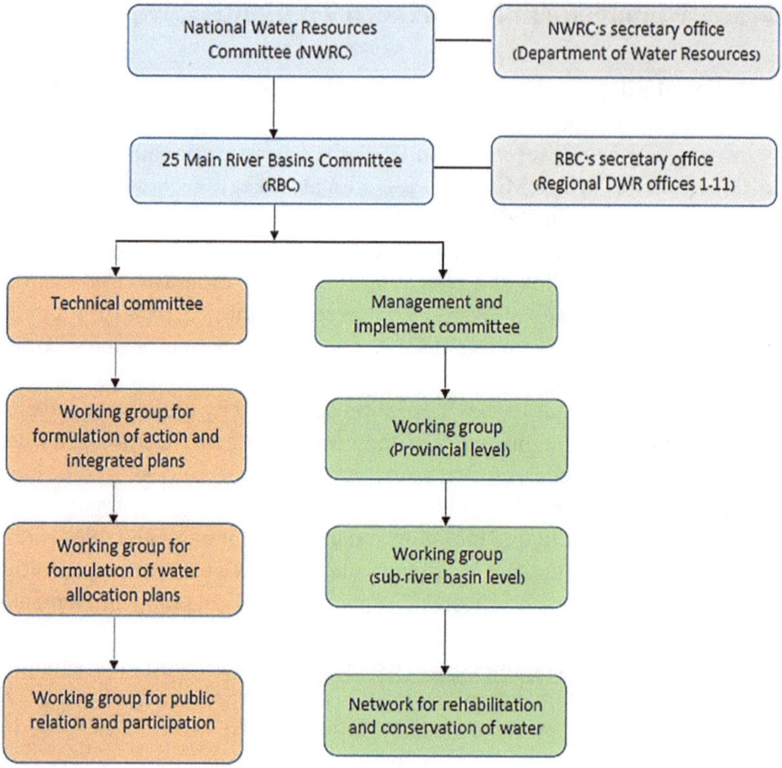

Chart 6.1 Organizational structures of NWRC and RBC (Source: Department of Water Resources (2006))

to the centralized public administration system of the country, each ministry also delegates its authorities and functions to its local agents to implement its policy by its own budget. For example, the Royal Irrigation Department (RID) under the Ministry of Agriculture and Cooperatives has offices in each province to implement its projects. This situation signifies that while the structure and line of command for water resource management is centralized, the resources for management, namely, budget and manpower, are fragmented. Thus, to achieve efficient and effective water management, inter-agency coordination is the key element. It should be noted that even with the well-coordination, there is a possibility that competition among agencies and duplication of projects may occur. Chart 6.2 displays the line of command and coordination of NWRC and other government agencies regarding water resource management.

In principle, each RBC performs its functions by formulating the water resources management plan in its area and coordinating the creation of an action plan by related agencies and the local government in line with the river basin's water resources management plan and budget frame (DWR 2014). According to its duties, the RBC organizes public hearings to compile information on local needs from local

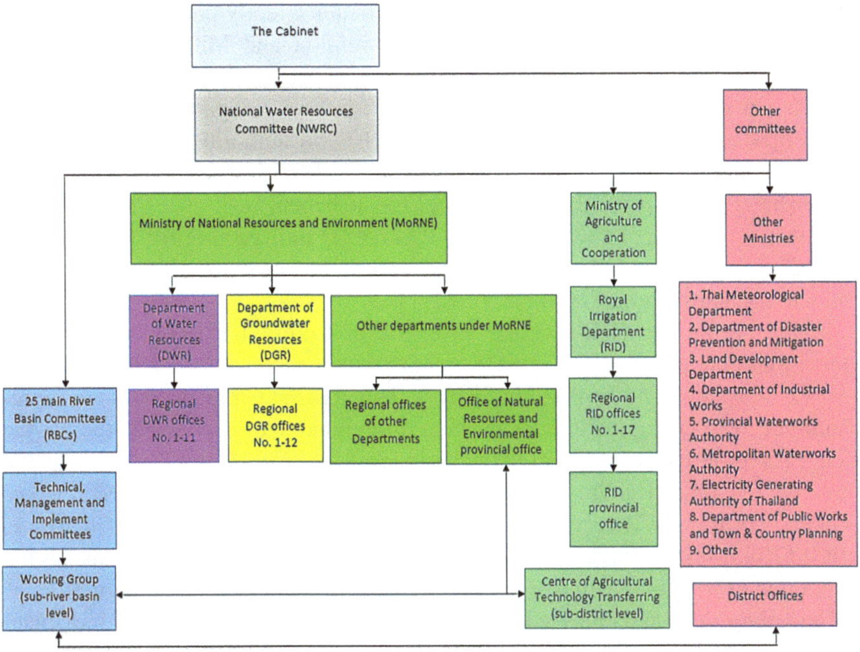

Chart 6.2 Line of command and coordination of NWRC and other government agencies for national water resource management (Source: Department of Water Resources (2006))

people and water users and drafts the river basin's development plan based on its findings and national development policy. However, in reality, RBC can only make a list of public needs and draft a plan since the authority given to it is limited due to lack of law or act, but only the Regulation of the Office of Prime Minister which has less authority than a law or act. In addition, the RBC in some basins has also been interfered by several government agencies that want it to include and implement projects of their choices into the plan instead of those based on information from public hearings.[7]

In terms of budget, although the planned projects of every government agency in each basin have to be stipulated in the development plan of that basin, and the RBC then submits it to the National Water Resources Committee to approve and forward to the Bureau of Budget for annual budget, some agencies have never proposed their projects to the RBC but submit to the central office of their departments or ministries with their acquired budget for implementing their projects independently from RBC. According to the interviews, "whether we propose our planned projects to the River Basin Committee or not, it does not matter because we are always allocated budget from the government and we can inform the River Basin Committee later when the projects are completed. Also, since the River Basin Committee has no its

[7] Interview with RBC of Kokand Mekong, April 2016.

own budget to allocate to our office, we have better directly proposed our project proposals to the Department of Disaster Prevention and Mitigation to get the budget,"[8] and "we directly propose project proposals to the Royal Irrigation Department not to the River Basin Committee, and we recognize the River Basin Committee as a project collector who totally has no authority to consider if the projects compiled should be implemented or allocated budget. Accordingly, the establishment of the River Basin Committee has no benefit at all since the Committee can function nothing."[9] These views signify another limitation of the RBC and the NWRC; they have no budget to encourage and no authority to regulate other government agencies to abide by their policies.

In the North, the Kok-Mekong River Basin Committee (RBC of Kok and Mekong) is appointed and assigned to manage water resource of Mekong River and its tributaries including Kok and Ing Rivers. This RBC's members consist of representatives from government agencies both at regional and local levels, the private sector, and groups of water users at the basin level. The Committee is chaired by the Governor of Chiang Rai Province, and the Director of the Regional Water Resources Office 1[10] of the DWR serves as its secretary. To function following its mandate, RBC of Kok-Mekong has also faced difficulties as mentioned above. This makes the committee to be only a feeble organization that is available but not important.

Regarding the stakeholder engagement, following the application of Integrated Water Resources Management (IWRM), NWRC and RBC have given priority to participation, particularly at the basin level. According to the Water Resource Management Strategies of DWR, "encouraging knowledge, understanding, and participatory process with civil society, basin based networks and organizations, local authorities, as well as other agencies involved in water resources management" is one of the implementation tactics of the DWR (DWR 2014, p. 70). However, as NWRC and RBC of which DWR serves as secretary have limited authority, budget, and manpower due to no law or act supported, its credibility of policy implementation is doubted. In addition, although NWRC and RBS have emphasized on the stakeholder engagement, their achievement is also limited due to the distrust and no participation of civil society in the area. As Neef mentioned, civic engagement in water management is low since civil society distrusts the government and believes that it cannot truly fulfill their needs (Neef 2008). This situation is also illustrated in case of the management of Kok and Ing Rivers, which some local NGOs, such as Rak Chiang Khong Group, do not avail their rights of participating in water management with the RBC due to the reason that the river basin development plan has never reflected appropriately their needs.[11]

In sum, in endeavoring to implement the IWRM, the Thai government has encountered several limitations and challenges. One of them is that the NWRC and RBC, which are the most important actors at the national and river basin levels that

[8] Interview with Office of the Disaster Prevention and Mitigation in Chiang Rai, April 2016.

[9] Interview with Regional Irrigation Office 2 in Chiang Rai, April 2016.

[10] This is the delegated authority of DWR at the regional level. There are 11 regional offices all over the country. Please see more details in Appendix 3.

[11] Interview with leaders of Rak Chiang Khong Group, April 2016.

has greatly influenced water resources management in Thailand, have no authority and resources to function effectively even though their structure and administrative body have existed for years. Without the law or act that grants the same degree of authority and resources as other agencies, the capacity, credibility, and achievement of these two agencies (NWRC and RBC) are doubted, resulting in the possibility of low degree of water governance, particularly in the river basin. The other challenge is the limited role of civil society and its participation in the decision-making process of water resource management. As water governance requires the active participation of civil society in all processes of policy decision-making, the passive role of civil society due to the past experience of mistrust between civil society and government officials has been an obstacle for practicing water governance in the river basin as well.

4 "People's Council of Ing River" and Water Allocation of Ing River

4.1 People's Council of Ing River

In 2011, the Network of Ing River Community and the Network for Natural Resources and Cultural Conservation in Mekong and Lanna Areas organized a meeting that all networks along the Ing River were invited to join. In this meeting, the idea of establishing the "People's Council of Ing River"was initiated and discussed. The concept of "People's Council" was proposed in order to transform the local movement and organization from a "community network" which was a loose cooperative form among local people and communities, to a more consolidated unit, a "council," with a permanent secretariat to facilitate the network's activities. Although the idea was widely discussed, there was no concrete action until 2013, when two more meetings were convened and the People's Council of Ing River became formally functional as a forum and a process for local people to participate in the Ing River development and management, particularly in the government's policy-making process related to the river. In addition, to support the livelihood and better standard of living of people along the river, the Council intends to extend and strengthen the existing network of acquiring knowledge and sharing it, as well as to expand the conservative areas along the river. The main principle of the Council is to provide opportunities and encourage local people to take part in the process of natural resources allocation and conservation along the Ing River in a fair and sustainable manner (Viset 2013).

As the Ing River flows through two provinces, Phayao and Chiang Rai, the Council was developed based on the existing community networks in those two provinces. In general, the river community network consists of representatives from several villages in communities along the river. Meetings and consultations are the main mechanisms of the community network. The basic function of all networks is to manage the daily water needs as well as monitor the government's development policy and projects that may affect the communities' livelihood. With agreement among the community network's committees, rules and practices for natural resource management in the communities, including allocation, utilization, conservation, and penalties, are set and enforced. There is also coordination among multiple networks of river communities (e.g., Network of Ing River Community, Network of Kok River Community, and Network of Sai River Community), and joint activities are conducted from time to time (Rakyuttitham 2000).

The Network of Ing River Community comprises several groups and networks in both Phayao and Chiang Rai. In 1993, people living along the upper Ing River in Phayao faced serious drought. When they realized that watershed degradation was the root cause of the problem, they introduced the Phayao Lake and Watershed Conservation Area in 1994 in order to restore the forest and the Ing River. The conservative area was looked after by the communities in that area—14 communities along 12 small tributaries that flow to the upper Ing River. Therefore, in 1994, with the demarcation of the watershed conservation area, the Phayao Lake and Watershed Conservative Group of 12 Tributaries was established. It has developed into the Network of Phayao Lake and Watershed of 12 Tributaries Group over the years. In addition, during 1995–1996, the groups' activities were extended to include the local fishery in the Phayao Lake.

Currently, the Network of Phayao Lake and Watershed of 12 Tributaries Group is an active member of the People's Council of Ing River together with other networks, namely, the Love Lao River Network, the Love Yuan River Network, the Network of Alternative Agriculture in Phayao, the Network of Traditional Fishery in Phayao Lake, the Network of Natural Resources of Phayao Province, and the Network of Mid Ing Rivers (Rakyuttitham 2000). It should be noted from the historical background and foundation of these networks that natural resources management, including allocation, utilization, and conservation, is their main concern, and their activities are conducted for the people's survival and sustainable livelihood. The activities, particularly those related to conservation, by nature, are not absolutely contradictory to the government's policy of reforestation and restoration of watersheds, nor do they cause serious conflict with government officials. To some extent, the government tries to cooperate with these networks and encourage them to participate in government-led activities.[12]

In contrast, in Chiang Rai, along the lower Ing River, the Rak Chiang Khong Group was formed in 1997 in order to protest against the government-initiated Kok-Ing-Nan Water Diversion Project, which was expected to heavily affect local people along the Ing River. The most critical issue regarding the project was the nonparticipatory process of the development plan. As residents of the area where the development project would be conducted, local people believed that they had the right to be informed of the plan in order to prepare for any changes. However, this expectation was not fulfilled, which disappointed them badly and led to suspicion and mistrust against the government's project and officials. In 1997–1998, the Rak Chiang Khong Group was able to gain support from several community networks in Phayao,

[12] Interview with government officials in Phayao, August 2015.

including the Network of Phayao Lake and Watershed of 12 Tributaries Group to conduct the protest (Viset and Boonserm 2004). This was the beginning of cooperation among the community networks of two provinces along the Ing River and was the foundation for the establishment of the People's Council of Ing River later. The key element that linked all the networks together was their approach toward river development focusing on holistic and integrated management of the river and water resources. Precisely speaking, as the river flows through the area without boundaries, and the water resources belong to those who live in the riparian area, they did not believe in having a river development plan separated by administrative boundaries or top-down management without the participation of local people living along the river. Their approach continues to be used and recently became the foundation for establishing the People's Council of Ing River.[13]

Due to the strong protest, the Kok-Ing-Nan Water Diversion Project was postponed. The success of the protest reflected the fact that the top-down approach was not accepted by local communities. It also illustrated the strength of community networks as a tool to negotiate with the government and demand for the right to protect community resources that belong to all (Viset and Boonserm 2004). In addition, it encouraged local people to form several networks in Chiang Rai, and in 2013, those networks, including the Network of Local People in Lower Ing River and the Network for Conservation of Lower Ing River, became members of the People's Council. Some networks were short-lived, such as, the Love Ing-Lao Rivers Group, while other networks remained active and developed into more solid organizations later, such as the Network for Social Life and Environmental Studies (Rakyuttitham 2000). It should be noted that, in comparison to the community networks in Phayao, the community networks in Chiang Rai are different in terms of their history, origin, purpose, and means of conducting activity. The main difference lies in the two groups' different views in the political dimension; namely, the networks and members in Chiang Rai are more policy-oriented with a desire and readiness to engage in the policy-making process and utilize political power or influence to achieve their objectives, while the networks and members in Phayao emphasize non-political activities and utilize the traditional way of life to conduct activities to achieve their objectives.[14] This difference, certainly, presents both opportunities and challenges for the People's Council of Ing River, whose membership includes all community networks from the two provinces. A great deal of compromise is, then, needed to conduct activities in the name of the Council. Concurrently, it is necessary to find common interests and shared burdens and benefits among the community networks to continue the Council.

The Council frequently carries out several activities: natural resources and cultural preservation, the establishment of the Foundation,[15] participation in the

[13] Interview with members of the Council in Phayao and Chiang Rai, August 2015.

[14] Observed by the author, leadership, namely, personality and charisma of networks' leaders in Chiang Rai and Phayao, may contribute to the difference between these two groups. However, this observation needs more study to confirm.

[15] Some members of the Council have an idea to raise fund by legalizing the Council through a formal registration as a "foundation."

policy-making process for development projects along the Ing River, and drafting of the Council's proposal for Ing River development and management. The strategic plan of action calls for the conservation and restoration of the watershed forest of Phayao Lake, demarcation and expansion of the fish conservation area, diversion of water from the Mekong River to the Ing River for agriculture, and a database to accumulate traditional knowledge (Viset 2013). Following an interview with members of the People's Council in Chiang Rai, the Council aims to advocate a development policy based on sustainable development and wants this policy to be included in the development policy of the province. Instead of being a reactive victim of the government's policy and plan, the Council seeks to initiate and propose its ideas to the government through formal and informal channels, including joining government- or province-led workshops or forums, or inviting officials to join its own workshops and forums.[16]

4.2 Different Views on Water Resource Allocation of Ing River

Regarding water resource management, the Council focuses on the development of areas along the Ing River and water utilization. The Ing River originated from Phi Pan Nam in Phayao province and flows through Phayao and Chiang Rai provinces upward to join the Mekong River at Chiang Khong District in Chiang Rai. Geographically, the Ing River is divided into three parts: the upper, middle, and lower parts. It covers a total area of 4773.34 square kilometers in the two provinces (River Basin Committee of Kok and Mekong River Basins 2012). It is 325 km long, and there are 23 small tributaries flowing in, which create a large area of wetland with plenty of natural resources including forests, wildlife, birds, fish, and plants. The water from the Ing River has contributed to the survival and livelihood of people in the riparian area as it is a source of food, daily consumption, and social values. Local people use the water for agriculture, husbandry, and washing and cleaning in daily life, as well as during worship according to traditional beliefs. Local people, who are both Thai and ethnic minorities, have earned a livelihood by utilizing the existing resources and concurrently preserving them based on the sufficiency philosophy and nature dependency (Yeunyong n.a.) (Fig. 6.1).

Recently, the degradation of rivers and watershed has become more serious. The River Basin Committee of Kok and Mekong River Basins (RBC of Kok-Mekong) identified the causes of degradation of the Ing River as the following: deforestation for agriculture expansion and soil erosion, water shortage and flood, and low quality of water due to chemical contamination from agriculture and residential areas. In addition, the RBC of Kok and Mekong indicated that the inefficiency and conflict related to water management occurred due to the different interests and perspectives of several actors and stakeholders. In general, competition for water for both agriculture and daily consumption among water users usually occur between

[16] Interview with the members of the Council, November 2016.

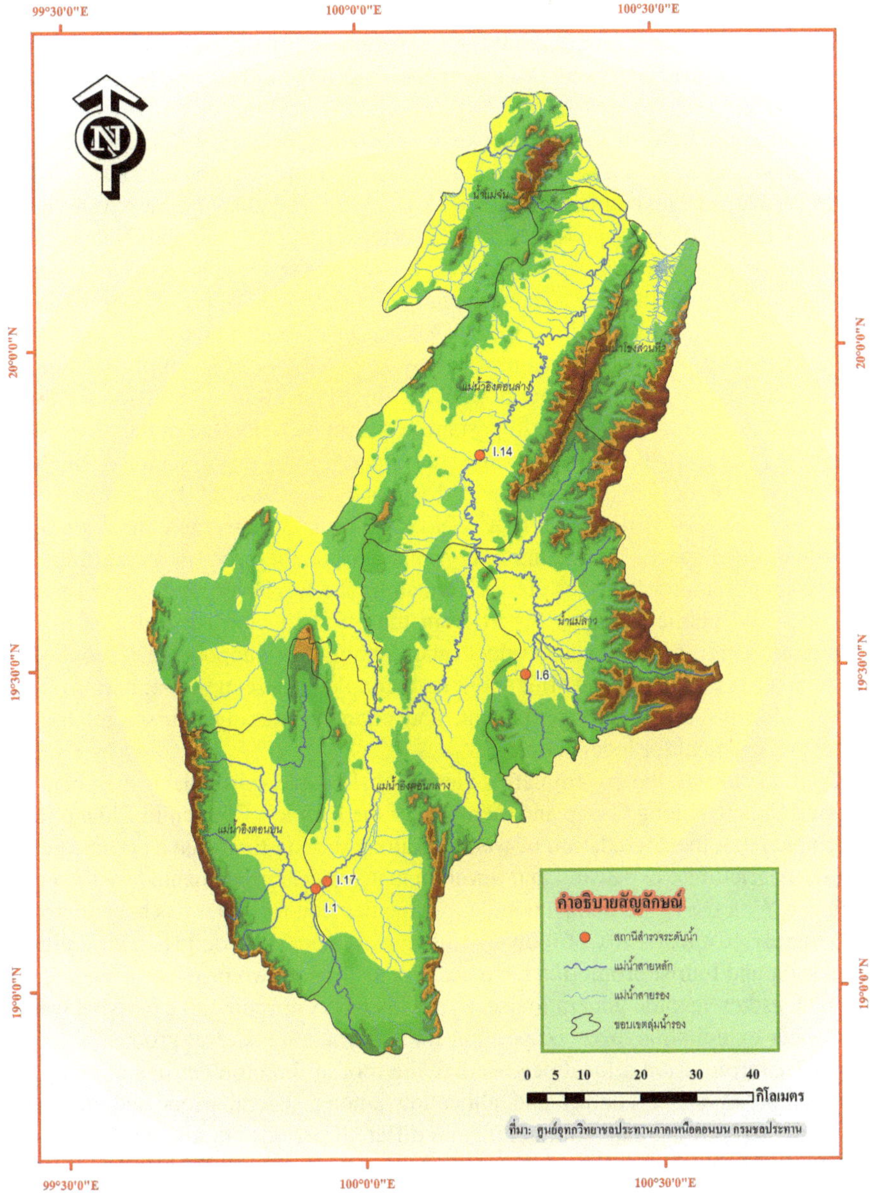

Fig. 6.1 Map of the Ing River Basin. (Source: Bureau of Water Management and Hydrology, Royal Irrigation Department https://www.hydro-1.net/Data/HD-01/1-07.html)

the communities upstream and downstream of the river. Concurrently, there existed conflicts between local communities and government officials regarding the water management and development projects in the area (RBC of Kok and Mekong River Basin 2012).

According to RBC of Kok and Mekong's document called the Kok and Mekong River Basin Integrated Development and Management Framework 2014–2016, RBC of Kok and Mekong has attempted to solve the problems of severe water degradation and shortage by giving priority to water supply management. Namely, the construction of a reservoir, wells, and a pipe system is recommended as a response to water shortage, while dykes, dams, and dredging are planned for flood prevention (RBC of Kok and Mekong River Basins 2012). The core idea that underlines these suggestions and plan is to manage water by increasing or controlling its amount through technology and construction. With the application of IWRM, RBC of Kokand Mekong recognized the importance of local participation in the process of water management, and concrete plans are supposed to be implemented during the 3-year period of the plan (RBC of Kok and Mekong River Basins 2012). Information from interviews with officials of RBC of Kokand Mekong illustrated their attempts to include local people and community networks in the process of drafting the development plan for water use in the area along the Ing River, by holding several meetings and public hearings. Apart from the effort to increase and control the amount of water, the officials conduct other activities to support local people for efficient water allocation and utilization; for example, they set up water user groups, conduct forums for dispute settlement, and organize training courses related to water management for local people.[17]

Contradictory to the government's views, local people, particularly members of the Rak Chiang Khong Group and members of the People's Council in Chiang Rai, have noted that the degradation of wetlands along the Ing River have occurred as a result of agriculture expansion and development projects implemented by the government. With the rising human population, the demand for water has been increasing sharply. Concurrently, following the government's guidelines, agricultural expansion and tourist promotion have been intensified in two provinces, since they are seen as key income generation sectors. This has contributed to excessive water use, water exploitation, and competition for water among several groups of water users. Therefore, from a local perspective, the root of the problem lies not only in the method of water control and allocation among several users and between upstream and downstream communities in different periods but also in the path of development that focuses on income growth and intensive utilization of natural resources (Yeungyong n.a.).

Due to the different perspectives regarding development direction and water utilization, local people and members of the People's Council in Chiang Rai found it necessary to negotiate with government officials. Data from interviews showed that in fact, in the past, members of the People's Council tried to voice their problems or propose their ideas through the participatory process initiated by RBC as well as

[17] Interview with officials of RBC of Kokand Mekong, August 2015 and November 2016.

other government agencies in the province. For example, some local people and members of the People's Council were appointed to sit in the RBC of the Kokand Mekong to give comments and draft the RBC's development plan. From RBC's perspective, these activities constituted local participation and stakeholder engagement. However, from local people's perspective, meetings without the mainstream ideas and comments about development policy and plan were meaningless and could not be categorized as "participation and engagement." In fact, they thought that they were only the "object" or "rubber stamp" of the government's policy and implementation plan. Therefore, some of them resigned at the end.[18]

The "understanding gap" between officials and local people or members of the People's Council in Chiang Rai reflects the need for RBC to reconsider its roles and activities to respond to the demands of local people. As Hirsh (2006) commented, in principle, RBCs have greater civil society and local participation than previous governance arrangements according to the Regulation of the Office of the Prime Minister on National Water Resource Management, but in practice, they are unable to apply that opportunity for water governance. At the same time, this "understanding gap" also provides the opportunity and space for civil society to be more active in the water governance process, particularly in the policy decision-making process, to secure efficient and effective water use for the Ing River.

5 "The Association of Chiang Saen Livable City" and "Kok River Basin Ecology Group" and Flood Prevention along the Kok River

5.1 *Overview of Two Local NGOs*

The Association of Chiang Saen Livable City is a group of local people living in Chiang Saen District, Chiang Rai province. Led by Mr. Suthep Lorsrithong, the Association officially registered as a formal entity with the Ministry of Interior of Thailand in 2005. Its members are local people in the area who are familiar with Mr. Lorsrithong and agreed to join the Association. Data from interviews did not show common ideas and interests among the Association's members in terms of water management or area development (Chiang Saen is the mouth of the Kok River and a historical city). According to the regulations, Mr. Lorsrithong has held official meetings and met the Association's members once a month to report the Association's activities, which are actually his activities on behalf of the Association. The main activity of the Association that Mr. Lorsrithong has carried out is participating in the government- or province-led meetings. These meetings cover a wide range of issues, such as, the development plan of the city, water management projects, and infrastructure development, as well as cultural and traditional events. Regarding water

[18] Interview with the members of the Council, August 2015.

management of the Kok River, it is interesting to note that Mr. Lorsrithong has been appointed as a member of RBC of Kokand Mekong while other members of the Association are rather passive.[19] Although the Association has existed for some time, its relations with other civil society organizations in Chiang Rai is rather limited due to the negative view that they share no common concerns or interests. For example, the Rak Chiang Khong Group pays attention to the conservation of Mekong giant catfish, but the Association of Chiang Saen Livable City is not interested in this activity, explaining that it is the job of the Department of Fisheries and not of local people and networks. Even though there are some cases where the Association shares the same view as other community networks, for example, China's threat of unexpected and unpredictable water release, which would affect people in the riparian areas of both the Kok and Ing Rivers, the role of the Association is still passive due to the lack of cooperation and support from people in the area owing to the fear of being noted as anti-government.[20] Compared with the Rak Chiang Khong Group located in Chiang Khong, the mouth of Ing River, the situation is quite different. It is also interesting to note that the Association, particularly Mr. Lorsrithong, does not participate in the activities of or cooperate with the People's Council of Ing River due to the perception that NGOs including the People's Council of Ing River and its activities are anti-government.

In terms of the relations with the government agencies, both central and provincial levels, Mr. Lorsrithong proudly explains that as the Chair of the Association of Chiang Saen Livable City, he represents one of the small number of civil society organizations invited to participate in meetings with the government agencies and be a part of a subcommittee for the RCB of Kokand Mekong.[21] This signifies the cordial relationship between the two sides. However, Mr. Lorsrithong also commented and admitted that although he has made an effort to voice his opinions to the government agencies, there is no guarantee that the government officials will recognize them. This is similar to the view expressed by the leader of the Rak Chiang Khong Group that actually the voice of local people is hardly heard by the government, particularly in the decision-making process.

Regarding water management of the Kok River, particularly flood prevention and early warning, although living at the mouth of the Kok River connecting to the Mekong River and often affected by flood, local people including members of the Association have been scarcely informed about the Mekong's situation by the related government agencies. Data from interviews illustrated their concerns; however, there has been no sign from the Association to send a signal to the government officials or cooperate with other community networks or civil society organizations, particularly the People's Council of Ing River, to raise this issue. Although they know about the telemetry and early warning system for flood prevention, local people do not find it appropriate to rely on the data and system implemented by government agencies.[22]

[19] Interview with members of the Association, April 2016.

[20] Interview with members of the Association, April 2016.

[21] Interview with the Chair of the Association, April 2016.

[22] Interview with members of the Association, April 2016.

Another group to be compared with the above civic organization in the same river basin is *The Kok River Basin Ecology Group*. This is one of the civil society organizations in Chiang Saen headed by Mr. Dusit Jitsook, a local scholar and farmer. Its activities focus on development and management of water usage in the Kok River Basin. Realizing the unsustainability of the government-led development project, particularly the construction of weir to manage the Kok River for multipurpose water utilization, Mr. Jitsook and local people living in the Kok River Basin established a group to jointly search for the method to restore the ecology and allocate water resources fairly to stakeholders in the area based on local wisdom.[23]

Regarding the cooperation with other civil society organizations in Chiang Rai, the Kok River Basin Ecology Group has cooperated with the Rak Chiang Khong Group due to the shared concerns and approach toward China's role in the region regarding Mekong River. Both groups realize the threat resulting from the Chinese dam construction upstream and the unexpected and unpredictable water release downstream for navigation by China, which causes unexpected floods in the downstream areas. To deal with this problem, it is necessary to cooperate and coordinate with both government agencies and NGOs. However, the experience of the Kok River Basin Ecology Group displayed that support from the government agencies was hard to find, while cooperation with other civil society organizations was more fruitful. The members of the Kok River Basin Ecology Group had also joined with the Rak Chiang Khong Group and coordinate with other networks in Northern Thailand to launch campaigns showing their opposition to hydropower development projects of upstream countries, including China and Lao PDR. It is interesting to note that the Kok River Basin Ecology Group does not cooperate with the Association of Chiang Saen Livable City, although they live and work in the same area, Chiang Saen, and Kok River.[24] This situation reflects the fragmentation of civil society organizations in the area, and it may lead to the weakness of these organizations in dealing with the government agencies. However, to be fair to these two groups, it should be observed that they have different standpoints and focuses. The Association of Chiang Saen Livable City concentrates on the development of Chiang Saen District, such as, maintaining and promoting their cultures and historical sites.[25] In contrast, the Kok River Basin Ecology Group aims to propose an alternative approach to development and water management in the Kok River Basin. With this difference of purpose, cooperation between the two groups seems to be very challenging.

For the relationship with the government, due to the different perspectives regarding development and water resources management, the Kok River Basin Ecology Group has rarely sought support from the government. In addition, it is seldom invited to participate in government- or province-led committees or projects; for example, the group's leader is not appointed as a member of RBC of Kokand Mekong, while the Chair of the Association of Chiang Saen Livable City is.

[23] Interview with members of the Kok River Basin Ecology Group, April 2016.

[24] Interview with members of the Kok River Basin Ecology Group, April 2016.

[25] Interview with members of the Association, April 2016.

However, the leader of the Kok River Basin Ecology Group has shared comments similar to those of the leader of the Rak Chiang Khong Group, suggesting that the participatory process initiated and practiced by the government agencies is meaningless if the people's proposed policy and plan are not included in the provincial development policy and plan.[26] In addition, the leader of the Kok River Basin Ecology Group has commented that although NGOs in Chiang Saen are strong and active, local people in the area are passive since their main concerns are issues related to their daily lives, for example, income generation, health, and education for their family members. This view is also shared by the Chair of the Association of Chiang Saen Livable City. The low degree of public awareness in Chiang Saen makes it difficult for NGOs to negotiate with government agencies for participation and engagement in the policy-making process due to limited support from the ground.

5.2 Flood Prevention in Chiang Saen: Integrated Cooperation and Local Participation?

The Kok River is a tributary of the Mekong River originated in Kengtung District, Myanmar. It meets the Mekong River near the Sop Kok Community in Chiang Saen District, Chiang Rai province, Thailand, where the confluence of the two is known as Sop Kok in the local language. The River is 128 km long in Myanmar and 157 km long in Thailand, having a total length of 285 km. Its basin in Thailand is around 7300.41 square kilometers in area, covering parts of Chiang Mai and Chiang Rai provinces. It is close to Myanmar and Lao PDR to the north, the Wang River Basin to the south, the Northern Mekong River Basin to the east, and the Ping River Basin as well as Myanmar to the west. There are three main geographic characteristics of the river basin as described by the Hydro and Agro Informatics Institute (2013): mountain, piedmont, and floodplain. Moreover, the Kok River Basin consists of four sub-basins: Mae Fang, where the basin area is 27.26% of the whole Kok River Basin; Mae Lao, 38.33%; Mae Saluay, 5.83%; and Lower Mae Kok, 28.28% (Fig. 6.2).

The Kok River has been affected by both natural and administrative problems, such as, floods, droughts, water scarcity, water quality, and water allocation and sharing. Furthermore, transboundary problems have threatened the river several times, particularly the downstream of the Kok River, where the surrounding area is easily flooded. The downstream flood occurs because of the Kok River itself and the relation between the Kok and Mekong Rivers. These flood types have normally been known as tributary[27] and combined[28] floods (Mekong River Commission

[26] Interview with the leader of the Kok River Basin Ecology Group, April 2016.

[27] Tributary flood occurs when the Kok River overflows its banks because of intense rainfall.

[28] Combined flood occurs as a result of the interaction between the Mekong River and Kok River when their water levels are high at the same time.

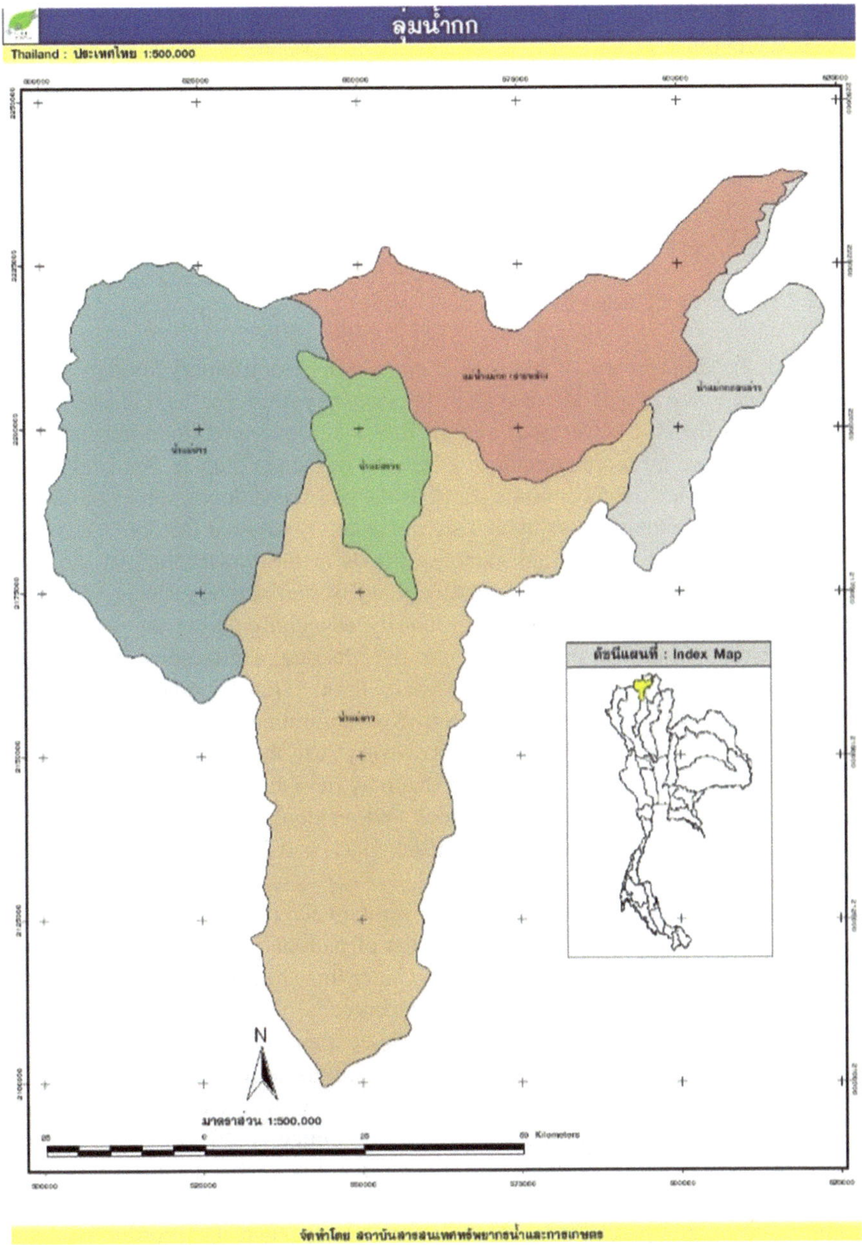

Fig. 6.2 Map of The Kok River Basin (Source: Hydro and Agro Informatics Institute)

Table 6.1 National and provincial agencies responsible for flood management

Central governmental agencies	Regional/local governmental agencies
1. National Water Resources Committee	1. Chiang Rai Provincial Office
2. Department of Water Resources	2. Chiang Rai Provincial Irrigation Office
3. Royal Irrigation Department	3. Kok and Mekong River Basin Committee
4. Department of Disaster Prevention and Mitigation	4. Disaster Prevention and Mitigation Provincial Office (Chiang Rai)
	5. Ban Saeo sub-district municipality
	6. Sob Kok Village Committee

Source: Compiled by the author

March 2012). Management of and mitigation measures for tributary floods might be easier as such flooding is affected only by factors within Thailand, such as, the water level of the Kok River and rainfall. However, combined flood management and mitigation are far more complicated because of factors outside Thailand's sovereignty, and this article concentrates only on the combined floods at the confluence of the Kok and Mekong Rivers, which are impacted by external factors.

The combined floods or transboundary floods at the downstream of the Kok River have been influenced by the relation of the Kok and Mekong Rivers. When the two rivers' water levels are high synchronously, especially during the wet season, the Kok River will not be able to flow into the Mekong, and the flow will reverse into the Kok River, causing a transboundary flood. Transboundary floods have impacted the downstream area where Sop Kok community is located for a long time. The statistical records show that almost every year, the community faces transboundary floods from the Mekong River that may devastate agricultural areas over 100 square kilometers, as well as livestock and residences.[29] It was recorded that severe transboundary floods occurred in 1966, 1971, and 2008, which were month-long flooding.

To manage and prevent flood, there are several involved agencies at both national and provincial levels. Table 6.1 shows the list of national and provincial agencies that are responsible for flood management. Under this situation, inter-agency coordination is very crucial, particularly during emergency.

At the national level, Thailand's water resource management strategy developed by the Royal Thai government in 2015 clearly stipulates that cooperation among related agencies and the participation of local people in water resource management are essential (Water Management and Policy Committee 2015). However, in practice, as mentioned earlier, each agency or department works independently with its own budget and is accountable to its home ministry. In addition, due to its own different purposes and goals, each agency is not aware about the coordination and duplication since it aims to fulfill its obligation mainly.[30]

---

[29] Interview with villagers of Ban Saeo sub-district municipality, April 2016.

[30] Interview with government official in Chiang Rai, August 2015 and April and November 2016, and in Bangkok in March and April 2016.

At the regional and provincial levels, although the RBC of Kokand Mekong is appointed and assigned for effective and integrated management of water resource including flood prevention, due to the limitation mentioned earlier, it is unable to accomplish its mission. Moreover, as there are other agencies in the area functioning for the same job, the duplication is inevitable because each agency carries out its activities in line with its own responsibilities assigned by its ministerial regulation, without recognizing how these activities should be integrated with those of other agencies.[31] To lucidly explain this issue, telemetry stations set up in the Kok River Basin by the Royal Irrigation Department and Chiang Rai Provincial Office to observe and forecast the water level of the Kok River and flood in its basin offer the best illustration of the lack of cooperation among governmental agencies.

Telemetry is defined as a highly automated communication process by which data are collected from instruments located in remote or inaccessible points and transmitted to receiving equipment for measurement, monitoring, display, and recoding. Telemetry is a water management system deployed by several agencies to generate data to be used in assessment and decision-making. It is used to manage water supplies for agriculture and to manage risk for early warning related to water quantity and quality. Data collection, data analysis, and data distribution are essential, and therefore, coordination is the key for management and utilization. There are more than 20 telemetry stations in the Kok River Basin installed by several central and local agencies including Royal Irrigation Department (RID), Department for Disaster Prevention and Mitigation (DDPM), and the Chiang Rai Provincial Office. Incongruously, two agencies have set up stations at the same place, never coordinated with each other to bridge their projects and thought that it was their responsibility to carry out their respective projects in accordance with their own terms of reference.[32] Data from interview suggested that in fact, every agency has considered the benefit to local people as a primary goal of its service, and the officials have conducted the projects with good intention. However, the centralized administrative system of the country has prevented them to realize about the incoordination and duplication that may happen afterward. In addition, although there are a number of telemetry stations installed by different agencies in the area, each agency utilizes its own information gathered by its own telemetry stations and rarely share information to other agencies or utilizes other agencies' information.[33] This reflects the inefficiency and ineffectiveness of information-sharing mechanism as well as telemetry stations. Finally, it should also be observed that almost all agencies installing telemetry stations had conducted the project as a "one-time activity" meaning that only the equipment was provided with no long-term maintenance cost. Then, the local authorities (municipal) or local people (village or community) have to bear the burden.[34]

---

[31] Interview with government officials in Chiang Rai in August 2015 and April and November 2016.

[32] Interview with government official of Chiang Rai Irrigation Project April 2016.

[33] Interview with government officials in Chiang Rai, August 2015 and April and November 2016.

[34] Interview with government officials in Chiang Rai, August 2015 and April and November 2016.

Regarding the local participation, data from interview displayed that civic participation has not been meaningful due to several reasons. From the government officials' point of view, they believe that flood prevention by telemetry system and stations is a responsibility of the government, not the local people. The local people are "object" or "beneficiaries" of the policy and project, who have no obligation to participate in the decision-making process and implementation. In addition, this project, by nature, requires technical knowledge which might be beyond local people's capacity; therefore, the government officials have rather monopolized the decision regarding the installation of stations and utilization of information.[35] On the other hand, from the local people's point of view, with the lack of knowledge and understanding about the system, it is difficult for them to utilize the information gathered by the stations for flood prevention. In addition, with the existence of social networks, local wisdom, and local leaders, they feel that it is more convenient, comfortable, and reliable to get information about the water current and its amount from their networks and leaders.[36] This situation illustrates that the local participation, although is encouraged, does not exist in reality. However, it should be noted that in consideration of local participation, the nature of the project does matter. Data from interview shows that while there was no local participation in telemetry system, local participation was high for the early warning system and evacuation rehearsal.[37]

The more critical issue regarding flood prevention is the combined flood caused by unexpected and unpredictable amount of water released from upstream for navigation. To respond to this situation, information sharing among involved agencies is critically important. Data from interview shows that information sharing among government agencies was limited and inefficient, while information distribution to local was less available as well. Although local communities, NGOs, and people in the area share the similar concerns, they seldom cooperate and voice their demands to the government both at provincial and national levels. The different standpoints regarding power and negotiation method is the main factor contributed to low degree of cooperation and no collective action among people and NGOs. The views from members of Sop Kok Village Committee and Association of Chiang Saen Livable City are similar, saying that the issue is too complicated and beyond their capacity to handle.[38] In contrast, the Kok River Basin Ecology Group has insisted that it is their rights to voice their concerns and demand for information from government agencies, as well as send a signal to the upstream countries to realize about the impacts of their policies and projects on downstream countries. This Group has cooperated closely with Rak Chiang Khong Group and the People's Council of Ing River. However, as mentioned earlier, without strong support from local people, this

[35] Interview with government officials in Chiang Rai, August 2015 and April and November 2016.

[36] Interview with villagers of Ban Saeo Sub-district Municipality, April 2016.

[37] Interview with villagers of Ban Saeo Sub-district Municipality, members of Sop Kok Village Committee, and members of Association of Chiang Saen Livable City, April 2016.

[38] Interview with villagers of Ban Saeo Sub-district Municipality, members of Sop Kok Village Committee, and members of Association of Chiang Saen Livable City, April 2016.

Group found it difficult to negotiate with the government not only in the case of transboundary flood but also the development of the river basin as a whole.[39]

6 Conclusion: The Reality of Life

It is believed that local participation and stakeholder engagement for water governance, while it is important and encouraged, is unintentionally prevented at the same time due to the institutional arrangement and centralized administrative system. However, some facts regarding key elements constructed to be "local participation and stakeholder engagement," such as roles of local NGOs, awareness of local people, and nature of the issues in the area, should be taken into consideration. The case studies of Mekong's tributaries in Northern Thailand shows that factors contributed to inefficient and ineffective water management of the Kok and Ing Rivers are from both government and civil society.

Many also believe that civil society, particularly local NGOs in the area where conflicts exist, is, in general, active and politically oriented. Table 6.2 illustrates similarities and differences among distinguished local NGOs in Chiang Rai where Kok and Ing River Basins are located. This fact signifies that local NGOs, although originated in the same area and witnessed the same problem, do not share the same interest and will not apply the same method to negotiate with the government. Unified NGOs may strengthen power for local participation; however, it is hardly accomplished due to several conditions, such as, different background and purpose of each NGO. In addition, it should also be noted that the strong NGOs and their active roles are not the prerequisite for successful negotiation. In fact, the perception of local people toward their status, rights, and power plays a crucial role in the process of local participation. This confirms the statements expressed by the leaders of both the People's Council of Ing River and Kok River Basin Ecology Group that without the public awareness, negotiation with either the national or neighboring countries' governments is almost impossible.

Concurrently, regarding the role of government officials, it is important to take note that only the institutional and organizational arrangement for local participation is not enough. The most important point is the perception of government officials regarding their authority and obligation and people's rights. As long as the government officials are unable to change their perceptions regarding local participation and transform their performance to recognize and promote "people's rights-based activities," their effort to encourage local participation will be only "ritual-based" which leads to no value for the local people.

Experiences in other continents, such as in Europe, offered a new paradigm for water resource management. Instead of an attempt to fix the state-centric institution and mechanism for water governance, "interactive governance" suggested by Edelenbos et al. may become the breakthrough for efficient and legitimate policy

[39] Interview with the leader and members of the Kok River Basin Ecology Group, April 2016.

Table 6.2 Comparisons of civic organizations

	The People's Council of Ing River	The Association of Chiang Saen Livable City	The Kok River Basin Ecology Group
Tributary (provinces)	The Ing River (Phayao and Chiang Rai provinces)	The Kok River (Chiang Rai province)	The Kok River (Chiang Rai province)
Year of establishment	2013	2005	2010
Aim and major activities	To promote sustainable development of Ing River Basin as well as Mekong River	To promote development of Chiang Saen District	To manage the Kok River Basin by local wisdom
			To promote sustainable development of Mekong River
Groups of networking	Wide linkage with several networks both in Chiang Rai and Phayao	No linkage with other networks	Rak Chiang Khong Group and other groups in the People's Council of Ing River
Relations with and attitude toward the government	Does not agree with government's view on local development	Good relationship with the government	Does not agree with the government's view of local development
	Does not participate in government's activities	Participate in the government's activities	Not be invited to participate in some government's activities
View toward local people	Need to be empowered	Concentrate only on their daily lives	Need to be empowered

Source: Compiled by the author

process. As Edelenbos et al. mentioned, interactive governance is an informal process with particular rules and roles that are different from the existing institutional representative system and is run parallel or prior to the formal institutions of negotiation and decision-making (Edelenbos et al. 2010, pp. 74). Termed as "civic initiatives" by Edelenbos and van Meerkerk, "it can be initiated by residents, social entrepreneurs, artists, and so on, as long as the initiative pursues a community purpose and not a direct business purpose…It is often assumed that the involvement of societal stakeholders can develop and enhance the efficiency, effectiveness, and legitimacy of decision making, implementation, and service delivery" (Edelenbos and van Meerkerk 2016, pp. 3–4). At the same time, as suggested by Edelenbos et al., the connections or interfaces between interactive governance and existing democratic institutions is also important for realizing the legitimate decision-making (Edelenbos et al. 2010). However, it should be noted that interactive governance has been developed based on European experiences of high degree of local participation and strong democratic institutions as a prerequisite. In Asia, particularly in Thailand and in Chiang Rai, this precondition does not exist. Then, the challenge at the local level in this country is more critical, and therefore, raising the awareness of people's right is the absolute requirement for interactive governance as well as legitimate policy decision-making and implementation.

Appendices

Appendix 1

25 RBCs in Thailand

1. Salween River Basin Committee
2. Ping River Basin Committee
3. Wang River Basin Committee
4. Kok and Northern Mekong River Basin Committee
5. Chao Phraya River Basin Committee
6. Sakae Krang River Basin Committee
7. Pa Sak River Basin Committee
8. Northeastern Kong River Basin Committee
9. Chi River Basin Committee
10. Mun River Basin Committee
11. Bang Pakong River Basin Committee
12. Prachin Buri River Basin Committee
13. Tonle Sap River Basin Committee
14. Eastern Coast River Basin Committee
15. Tha Chin River Basin Committee
16. Mae Klong River Basin Committee
17. Phetchaburi River Basin Committee
18. Western Coast River Basin Committee
19. Eastern South Coast River Basin Committee
20. Songkhla Lake River Basin Committee
21. Pattani River Basin Committee
22. Yom River Basin Committee
23. Nan River Basin Committee
24. Tapi River Basin Committee
25. Western South Coast River Basin Committee

Appendix 2

Agencies Involved with Water Resources Management (DWR 2014)

1. Office of the Prime Minister
   1.1 Office of the National Economic and Social Development Board
   1.2 Budget Bureau
2. Ministry of Agriculture and Cooperatives
   2.1 Office of the Permanent Secretary/Bureau of Rainmaking and Agricultural Aviation
   2.2 Royal Irrigation Department
   2.3 Fishery Department
   2.4 Land Development Department
   2.5 The Cooperatives Promotion Department
   2.6 Department of Agricultural Extension
   2.7 Agricultural Land Reform Office
   2.8 Office of Agricultural Economics
3. Ministry of Transportation
   3.1 Marine Department
4. Ministry of Interior
   4.1 Department of Provincial Administration
   4.2 Department of Disaster Prevention and Mitigation
   4.3 Department of Public Works and Town Planning
   4.4 Department of Local Administration
5. Ministry of Information and Communication Technology
   5.1 The Meteorological Department
6. Ministry of Natural Resources and Environment
   6.1 Office of Natural Resources and Environment Policy and Plan
   6.2 Department of Pollution Control
   6.3 Department of Environment Quality Promotion
   6.4 Department of Water Resources
   6.5 Forestry Department
   6.6 Department of Groundwater Resources
   6.7 Department of National Park, Wildlife, and Plants
   6.8 Department of Marine and Coastal Resources
7. Ministry of Industry
   7.1 Department of Industrial Factory

8. Ministry of Defense
   8.1 Armed Force Development Command, Royal Thai Armed Force
   8.2 Hydrographics Department, Royal Thai Navy
9. Ministry of Energy
   9.1 Department of Alternative Energy Development and Efficiency
10. Bangkok Metropolis Authority
    10.1 Department of Drainage and Sewerage
11. Independent Public Agency
    11.1 National Research Council of Thailand
12. State Enterprise
    12.1 Electricity Generating Authority of Thailand
    12.2 Metropolitan Water Supply Authority
    12.3 Provincial Water Supply Authority
    12.4 Industrial Estate Authority of Thailand

Appendix 3

11 Regional Offices of DWR

No.	Office	River Basin	Province
1.	Regional Office 1 in Lampang Province	1.1 Salween River Basin	Mae Hong Son Tak Chiang Mai
		1.2 Ping River Basin	Chiang Mai Lumphun Tak Kamphaeng Phet Nakhon Sawan
		1.3 Wang River Basin	Lampang Tak Chiang Rai
		1.4 Kok and Northern Mekong River Basin	Chiang Rai Phayao Chiang Mai

(continued)

No.	Office	River Basin	Province
2.	Regional Office 2 in Saraburi Province	2.1 Chao Phraya River Basin	Nakhon Sawan Chai Nat Sing Buri Ang Thong Phra Nakhon Si Ayutthaya Pathum Thani Nonthaburi Bangkok Samut Prakarn Lopburi Saraburi Phetchabun Kamphaeng Phet
		2.2 Sakae Krang River Basin	Uthai Thani Nakhon Sawan Kamphaeng Phet
		2.3 Pa Sak River Basin	Phetchabun Lopburi Saraburi Loei Chaiyabhum Phra Nakhon Si Ayutthaya
3.	Regional Office 3 in Udon Thani Province	3.1 Northeastern Kong River Basin	Loei Udon Thani Nongkai Bungkan Sakon Nakhon Nakhon Phanom Mukdahan Nong BuaLumphu Amnat Charoen Ubon Ratchathani
4.	Regional Office 4 in Khon Kaen Province	4.1 Chi River Basin	Chaiyabhum Khon Kaen Maha Sarakham Kalasin Roi Et Yasothon Loei Phetchabun Nong Bua Lumphu Udon Thani Nakhon Ratchasima Mukdahan Sisaket Ubon Ratchathani

(continued)

No.	Office	River Basin	Province
5.	Regional Office 5 in Nakhon Ratchasima Province	5.1 Mun River Basin	Nakhon Ratchasima Buriram Surin Sisaket Ubon Ratchathani Amnat Charoen Chaiyabhum Khon Kaen Maha Sarakham Roi Et Yasothon
6.	Regional Office 6 in Prachinburi Province	6.1 Bang Pakong River Basin	Chachoengsao Nakhon Nayok Prachinburi Sa Kaeo Chonburi
		6.2 Prachinburi River Basin	Prachinburi Chachoengsao Nakhon Nayok Sa Kaeo Chonburi
		6.3 Tonle Sap River Basin	Sa Kaeo Prachinburi Chanthaburi
		6.4 Eastern Coast River Basin	Chonburi Rayong Chanthaburi Trat
7.	Regional Office 7 in Ratchaburi Province	7.1 Tha Chin River Basin	Suphan Buri Nakhon Pathom Samut Sakhon Chai Nat Uthai Thani
		7.2 Mae Klong River Basin	Kanchanaburi Ratchaburi Samut Songkhram Tak Uthai Thani Suphan Buri Nakhon Pathom
		7.3 Phetchaburi River Basin	Phetchaburi Ratchaburi Samut Songkhram Prachuap Khiri Khan
		7.4 Western Coast River Basin	Prachuap Khiri Khan Phetchaburi Chumphon

(continued)

No.	Office	River Basin	Province
8.	Regional Office 8 in Songkhla Province	8.1 Eastern South Coast River Basin	Chumphon Nakhon Si Thammsarat Narathiwat Surat Thani Songkhla Pattani Yala
		8.2 Songkhla Lake River Basin	Songkhla Phatthalung Nakhon Si Thammsarat
		8.3 Pattani River Basin	Yala Pattani Songkhla
9.	Regional Office 9 in Phitsanulok Province	9.1 Yom River Basin	Phrae Sokhothai Phayao Lampang Nan Phitsanulok Tak Phichit Kamphaeng Phet Nakhon Sawan
		9.2 Nan River Basin	Nan Uttaradit Phitsanulok Phichit Phetchabun Nakhon Sawan
10.	Regional Office 10 in Surat Thani Province	10.1 Tapi River Basin	Surat Thani Nakhon Si Thammsarat Krabi Trang
		10.2 Western South Coast River Basin	Ranong Phang Nga Phuket Krabi Trang Satun Nakhon Si Thammsarat
11.	Regional Office 11 in Ubon Ratchathani Province	11.1 Lower Kong Chi Mun River Basins	Ubon Ratchathani Amnat Charoen Mukhahan Yasothorn

References

Badenoch, Nathan, Kate Lazarus, Bernadette P. Resurreccion, and Nga Dao. 2012. Water Governance and Water Rights in the Mekong Region. In *Water Rights and Social Justice in the Mekong Region*, ed. Kate Lazarus, Bernadette P. Resurreccion, Nga Dao, and Nathan Badenoch, 1–15. London: Earthscan.

Department of Water Resources (DWR). 2014. *Water Management*. Bangkok: Bureau of Public Promotion and Coordination, DWR.

———. 2006. *Water Management*. Bangkok: Bureau of Public Promotion and Coordination, DWR.

Edelenbos, Jurian and Ingmar van Meerkerk (2016). Introduction: Three Reflecting Perspectives on Interactive Governance in Jurian Edelenbos and Ingmar van Meerkerk. In *Critical Reflections on Interactive Governance: Self-Organization and Participation in Public Governance*, 1–28. Cheltenham: Edward Elgar Publishing Inc., Elgar online: https://doi.org/10.4337/9781783479078

Edelenbos, Jurian, Nienke van Schie, and Lasse Gerrits. (2010). Organizing Interfaces Between Government Institutions and Interactive Governance. *Policy Science* 43: 73–94. Springer online: https://doi.org/10.1007/s11077-009-9086-2

Hirsh, Philip. 2006. Water Governance Reform and Catchment Management in the Mekong Region. *The Journal of Environment & Development* 15 (2): 184–201.

Imamura, Masao. 2007. Introduction: Water Governance in the Mekong Region. In *Democratizing Water Governance in the Mekong Region*, ed. Louis Lebel, John Dore, Rajesh Daniel, and Yang SaingKoma, 1–8. Chiang Mai: Mekong Press.

Mekong River Commission. (2012, March). *Flood in the Lower Mekong Basin*. Working Paper 2011–2015: The Impact & Management of Floods & Droughts in the Lower Mekong Basin & the Implications of Possible Climate Change. p. i.

Neef, Andreas. 2008. Lost in Translation: The Participatory Imperative and Local Water Governance in North Thailand and Southwest Germany. *Water Alternatives* 1 (1): 89–110.

OECD. 2015. *The OECD Principles on Water Governance*. Center for Entrepreneurship, SMEs, Regions and Cities.

Rakyuttitham, Atchara. (2000). Community Network: Resources Management in Watershed Area by Local Participation. Document for the Seminar titled "Resources Management in the Watershed Area by Local Participation". 26–27 October, 2000, Chiang Mai.

River Basin Committee of Kok and Mekong River Basins (RBC of Kok and Mekong). 2012. *The Kok and Mekong River Basins Integrated Development and Management Framework 2014–2016*. Bangkok: RBC Secretariat.

Rogers, Peter, and Alan W. Hall. (2003). *Effective Water Governance*. Global Water Partnership, Technical Committee (TEC) Background Papers No. 7, February 2003.

Santasombat, Yos. 2011. *The River of Life*. Chiang Mai: Mekong Press.

Viset, Sahatthaya. (2013). Minutes of the People Council of Ing River Meeting on 10 June, 2013, Phayao.

Viset, Sahatthaya, and Nikom Boonserm. 2004. *Water Management by Community*. Bangkok: Chulalongkorn University Press.

Water Management and Policy Committee. (2015). *The Strategic Plan on Thailand's Water Resources Management*. Bangkok: Water Management and Policy Committee.

Yeunyong, Pathara. n.a. *Ing Watershed Management*. Phayao: Project on Development for Phayao supported by GEF/UNDP.

Website

http://www2.haii.or.th/wiki/index.php/%E0%B8%AA%E0%B8%A0%E0%B8%B2%E0%B8%9E%E0%B8%A0%E0%B8%B9%E0%B8%A1%E0%B8%B4%E0%B8%9B%E0%B8%A3%E0%B8%B0%E0%B9%80%E0%B8%97%E0%B8%A8%E0%B8%A5%E0%B8%B8%E0%B9%88%E0%B8%A1%E0%B8%99%E0%B9%89%E0%B8%B3%E0%B8%81%E0%B8%81

http://www2.haii.or.th/wiki/index.php/%E0%B9%84%E0%B8%9F%E0%B8%A5%E0%B9%8C:03_%E0%B8%A5%E0%B8%B8%E0%B9%88%E0%B8%A1%E0%B8%99%E0%B9%89%E0%B8%B3%E0%B8%81%E0%B8%81.jpg

www.oecd.org/env/watergovernanceprogramme.htm

www.watergovernance.org/water-governance/

www.watergoveranace.org/governance/what-is-water-governance/

Siriporn Wajjwalku is an Associate Professor of the Faculty of Political Science, Thammasat University, Bangkok, Thailand. In addition, she is also the Acting Dean of the School of Social Innovation, Mae Fah Luang University, Chiang Rai, Thailand. Her research interests are in the field of international relations and international development focusing on transboundary issues in Mekong Subregion such as food security, water governance, and disaster risk reduction. As the Project Coordinator of Thailand Research Fund, she has conducted several collaborative projects with both Thai and foreign partners.

Chapter 7
Interactive Governance for Sustainable Resource Use and Environmental Management: A Case Study of *Yaman ng Lawa* Initiative in the Laguna Lake Watershed, Philippines

Tadayoshi Masuda

Abstract This chapter deals with the watershed and water resource management of Laguna Lake in the Philippines. Laguna Lake is the largest freshwater lake in the Philippines and is located next to the capital, Metropolitan Manila. Due to urbanization and industrialization, the lake's water quality is deteriorating, and water resource management is now urgently required. This chapter reviews the Philippines' water governance and its history and framework and then examines the activities and roles of stakeholders for Laguna Lake. This chapter also focuses on the Laguna Lake Development Authority and the role of the Yaman ng Lawa Initiative in local community activities for managing and using the lake and its watersheds sustainably. This chapter discusses how community mechanisms can produce social capital by interactive governance in the lake basin.

Keywords Interactive governance · Water resource use and environmental management · *Yaman ng Lawa* · Laguna Lake watershed · Philippines

1 Introduction

Interactive governance, advocated by Torfing et al. (2012), raised a new paradigm for governance study and advanced it, especially in the area of watershed and water resources management. However, this framework of idea was constructed conceptually by considering water governance in Europe (e.g., Edelenbos et al. 2010), and there is a concern regarding its universality. Is interactive governance adaptable and acceptable in Asia where the social and natural conditions are different from those of Europe? Chapter 7 deals with watershed and water resources management

T. Masuda (✉)
Department of Agricultural Science, Faculty of Agriculture, Kindai University, Nara, Japan

K. Otsuka (ed.), *Interactive Approaches to Water Governance in Asia*,
https://doi.org/10.1007/978-981-13-2399-7_7

through a case study of *Yaman ng Lawa* ("Blessings of the Lake" *in Tagalog*) Initiative in the Laguna Lake, Philippines. Laguna Lake is the largest freshwater lake in the Philippines and is located next to Metropolitan Manila, the nation's capital. Due to urbanization and industrialization, the lake water quality is a concern. In addition, the lake's water resources management is an urgent issue. First, this chapter summarizes the Philippines' water governance as well as its history and framework following Rola et al. (2015). Second, stakeholders for Laguna Lake are introduced, and their activities and roles are arranged, focusing on the Laguna Lake Development Authority (LLDA). Third, "Yaman ng Lawa," local community activities with objectives to manage and utilize Laguna Lake and its watersheds sustainably are introduced. Fourth, this chapter discusses community mechanisms to produce social capital (Hayami 2009) and interactive governance, also referring to Ostrom (1990) as well as some case studies in Southeast Asia.

2 Water Governance in the Philippines

Rola et al. (2015) summarize water governance in the Philippines (Table 7.1) and at least seven legal frameworks that govern the water sector in the Philippines:

Table 7.1 Evolution and drivers of water governance reforms in the Philippines

Period	Enabling law	Implementing agencies	Drivers of reform
Post war, 1946–1955	Customary rule	Community-based local authorities: National Waterworks and Sewerage Authority (NAEASA)	Spirit of collectivism
	Constitution; water is a right		Economic development; recovery from war damage
Pre-martial law, 1955–1971	Republic Act 6234	Metropolitan Waterworks and Sewerage System (MWSS, formerly NAWASA)	Economic development; demographic growth
Martial law, 1972–1985	PD 198 or the Provincial Water Utilities Act of 1973	Local Water Utilities Authority (LWUA)	National-local partnership
	PD 1067 or the Water Code of the Philippines of 1976	National Water Resources Board (NWRB, formerly National Water Resources Council – NWRC): Rural Waterworks Development Corporation (RWDC)	Need for water right assignment; solutions for water conflict; international pressure
Post-martial law, 1986-present	Local Government Code of 1991; Water Crisis Act 1997; RA 9275 or the Clean Water Act 2004	Local government units; privatization of the MWSS; Department of Public Works and Highways (DPWH); water quality and pollution control	Market-based mechanisms such as pollution permits: Co-management; privatization drought; decentralization partnership

Source: Rola et al. 2015

(i) Presidential Decree 1067 Water Code (1976)
(ii) Presidential Decree 198 Provincial Water Utilities Act (1973)
(iii) Presidential Decree 522 (1974)
(iv) Republic Act 7586 National Integrated Protected Areas System Act (1992)
(v) Republic Act 8041 National Water Crisis Act (1995)
(vi) Republic Act 8371 Indigenous Peoples Rights Act (1997)
(vii) Republic Act 9275 Clean Water Act (2004)

According to Hall et al. (2014), the seven laws cover the following:

1. The legal treatment of water resources (ownership, rights, and distinctions between types of surface or groundwater and sectors/users)
2. Property rights (including private rights granted to individuals versus the collective; whether rights can be leased, transferred, or recalled by the granting authority) and their enforcement
3. Legalized inter-sectoral prioritization and the basis for prioritization
4. Legal linkages between the land and surface water and between the land and forest and/or the environment
5. Intergovernmental responsibility for the water law, meaning that many agencies promote the law
6. Combining the water law with other laws on land, forest, and the environment and for water planning and development
7. Favoring the private sector and NGO participation in water planning and development
8. Openness to market solutions

Rola et al. (2016) also provide an empirical basis for local water governance reforms in the Philippines using primary data from 299 water managers representing water districts, community-based water organizations, irrigators' associations, and local government water providers. Their survey results showed that local water organizations are subject to various forms of laws and regulations; there are no set bases for water price determination; and water administration setup varies from formal to loosely informal structures. Recommendations are offered toward responsive local water governance arrangements.

3 Laguna Lake and Its Lakeshore Region[1]

Laguna Lake is blessed with natural resources that have market and nonmarket values. Because Metropolitan Manila[2] is included as lakeshore, the lake is utilized in various ways. Fishery production was 150,957 metric tons in 2012. Aquaculture

[1] Refer to the LLDA web site: http://www.llda.gov.ph/ See Appendix for outline of the lake.

[2] Taguig City, Pasay City, Pasig City, Pateros, Muntinlupa City, Marikina City, Caloocan City, City of Manila, and Quezon City

produced 63,490 metric tons in 2012. Municipal fisheries produced 87,467 metric tons in 2012. As business establishments, there are approximately 500,000 (estimated by LLDA) large, medium, and small enterprises. Regarding electricity, the total dependable capacity of the three plants was 758 megawatts as of 30 June 2015: Kalayaan Pumped-Storage Hydroelectric Power Plant, 28 megawatts; Caliraya Hydroelectric Power Plant, 28 megawatts; and Botocan Hydroelectric Power Plant, 10 megawatts. Regarding industrial cooling, 2.04 billion square meters of lake water is used annually. Regarding agriculture, the potential total irrigation area is 32,684, estimated as of December 2013. Recreation usage includes fishing, boating, sailing, and swimming (Fig. 7.1). The domestic water supply is provided by Maynilad Water Services, Inc.

Laguna Lake existing usages are shown below.

Fisheries

The lake's most dominant use is fisheries. It is estimated that the lake yielded 37,000 to 47,000 metric tons of fish from 1997 to 2000 – both from fish pens and open fisheries. Crustaceans and mollusks declined in yield from 1927 to 574 metric tons and from 101,101 to 85,363 metric tons, respectively, during the same period. With the introduction of fish pen culture technology in 1970, the boom in lake fishery production also led to social conflicts in the region. Large-scale fish pens have threatened the economic conditions of fishermen dependent upon the lake as an open-water fishery. To address the issue, a revised Lake Fishery Zoning and Management Plan was implemented in 1996.

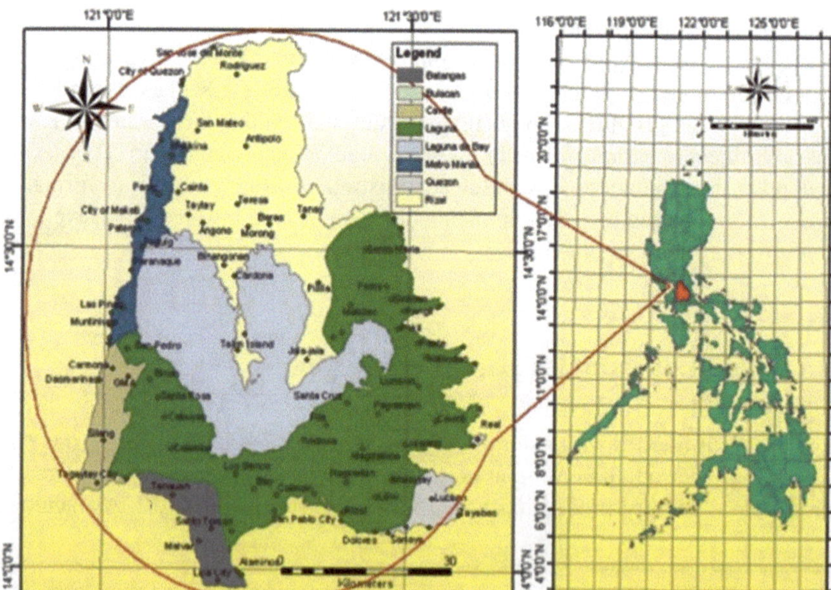

Fig. 7.1 Laguna Lake, Philippines (Source: RIHN)

Transport Route

LLDA set navigational routes for lake travelers and users to avoid conflicts with fish cultivators, fish pen owners, and free-range fishermen. More than 5000 motorized and non-motorized watercrafts operate as a means of transportation for lakeshore communities. In addition, there are 23 barges operating in the lake to transport an average of 75,640 barrels of oil and oil products to various supply depots daily.

Board Resolution No. 66 of 1998 approved the navigational routes for the ferry system in Laguna de Bay.

- Primary lane, a 500-meter-wide route going to Los Baños and Pagsanjan, Laguna
- Secondary lane, a 400-meter-wide route that provides access to other municipalities around the lake including Talim Island
- Tertiary lane, a 200-meter-wide route that leads to the existing landing areas in the municipalities

Flood Water Reservoir

The lake also serves as a reservoir for floodwater to save Metropolitan Manila from flooding. The Manggahan Floodway was constructed to divert floodwaters from Marikina River into the lake. The Napindan control station regulates the outflow of excess lake waters and minimizes the inflow of saline water and pollution from the Pasig River.

Power Generation

Laguna de Bay is also used for power generation; three power plants are located in the region. A pumped-storage hydroelectric power station is operated in Kalayaan, Laguna. Water is pumped up to the Caliraya reservoir to generate about 300 megawatts of electricity. Efforts are underway to increase this capacity to 600 megawatts.

Recreation

Laguna de Bay is also known for recreational activities. Although classified only for non-contact recreation such as fishing, boating, and sailing, the lake is used for swimming in some areas. Lakeshore resorts near Mt. Makiling extract hot spring waters for health spas and beauty treatment. LLDA also has ecotourism projects through the LISCOP component 1 project with six eco-enhancement projects in the Rizal and Laguna Provinces, namely:

A. Daranak Falls Eco-Tourism Project
B. Panguil River Eco-Park (Ambon-Ambon Falls)
C. Pakil Eco-Tourism Project
D. Majayjay Eco-Tourism Project (Taytay Falls)
E. Cavinti Eco-Tourism Project (Magdapio Falls)
F. Siniloan Eco-Tourism Project (Buruwisan Falls)

Irrigation
Laguna de Bay also provides sufficient water for farmlands in the lake region. Studies are being done to further improve the use of lake water for agriculture.

Industrial Cooling
The lake is being used as a source of industrial cooling water. Major users include the National Power Corporation for the Kalayaan Pumped-Storage Hydroelectric Power Plant and the Philippine Petroleum Corporation Refinery in Pililla, Rizal. KEPHILCO Malaya Thermal Power Plant also draws water from the lake for cooling. Cooling water is recycled back into the lake resulting in thermal pollution of about a 20-degree-centigrade rise in temperature near the discharge points.

Water Sink
The lake also serves as a huge waste sink for solid and liquid waste coming from households, cropland areas, industries, and livestock and poultry production as well as fishery activities. Not a single municipality is equipped with a sewerage system, so pollution is carried as surface runoff through the lake's sub-basins. In addition, polluted waters from the Marikina and Pasig Rivers also flow into the lake.

Source of Potable Water
The lake is now a major resource that supplies the domestic water supply requirement of the nearby towns along its coastline. This is supported by Board Resolution No. 338, series of 2007: "Declaring the Establishment and Operationalization of Water Permitting, Registration and Monitoring System for the Extraction of the Lake Waters of Laguna de Bay and other Bodies of Water within the Laguna de Bay Region and for Other Purposes" and backed-up by the approval of the Implementing Rules and Regulations for Permitting Surface Water Abstraction in 2008. Water permit for Manila Water Services Inc. (MWSI) to abstract a maximum volume of 300,000 cubic meters per day was approved in July 2009.

4 Laguna Lake Stakeholders: LLDA, Local Government Units, and Barangays

4.1 Laguna Lake Development Authority (LLDA)[3]

The Laguna Lake Development Authority or LLDA was organized in 1970 by virtue of Republic Act No. 4850 as a quasi-government agency with regulatory and proprietary functions. Through Presidential Decree 813 in 1975 and Executive Order 927 in 1983, its powers and functions were further strengthened to include environmental protection and jurisdiction over the lake basin's surface water. In

[3] Refer to the LLDA web site: http://www.llda.gov.ph/

1993, through the Executive Order 149, the administrative supervision over LLDA was transferred from the Office of the President to the Department of Environment and Natural Resources (DENR).

The LLDA's vision is that by 2020, the Laguna de Bay Basin will be transformed as the focal center for sustainable development through sound ecological governance. Its mission is Integrated Water Resource Management in the Laguna de Bay Region, with a clear focus on preserving ecological integrity and promoting sustainable economic growth.

As an environmental management program, the LLDA puts into action systems and programs such as the Environmental User Fee System (EUFS), the Public Disclosure Program (PDP), the Appropriation and Utilization of Surface Waters (Surface Waters), and the Compliance Assistance Centers (CAC) for the management of the lake waters and tributaries that flow into the Lake to ensure their proper use and maintenance for the sustainability of the ecosystem.

User Fee System (EUFS)
The EUFS is a market-based instrument that applies the "polluter pays principle (PPP)." It covers all enterprises within the lake region and requires them to obtain a Discharge Permit (DP) – a legal authorization for the enterprise to discharge their treated wastewater meeting the set environmental standards into the tributaries within the Laguna de Bay Region.

Public Disclosure Program (PDP)
In addition to the EUFS, the LLDA under the PDP assesses the environmental performance of regulated establishments and local government units (LGUs) and discloses the results thereof to the public. This mechanism is intended to motivate the industrial sector and LGUs to reduce their pollution.

Surface Waters Program
Under the Surface Waters program, Water Permitting, Registration, and Monitoring programs are established and operated for extracting waters that are naturally open to the atmosphere, such as rivers, lakes, and streams within the Laguna de Bay Region.

Compliance Assistance Center (CAC)
To encourage and improve compliance with environmental regulations and standards, the CACs connect the LLDA with regulated establishments initially in the hog/poultry farms and slaughterhouse sectors. The CACs also provide clear and consistent information to these establishments regarding environmental laws and regulations.

To abate the further degradation of the Laguna de Bay watershed, the LLDA instituted various programs and strategies for reforestation of the denuded watersheds. Part of these programs and strategies are the Laguna de Bay Community Carbon Finance Project (LDBCCFP) and participation in the National Greening Program (NGP), the River Rehabilitation Program, and the Shoreland Management Program.

Laguna de Bay Community Carbon Finance Project (LDBCCFP)
Back to back with the LISCOP project, the LDBCCFP is an initiative that aims to reduce carbon emissions through local government projects and activities funded by LLDA through the World Bank. LDBCCFP engages communities to implement solid waste management, composting facilities, and reforestation projects. Under the LISCOP project, the LLDA likewise finances construction of infrastructure by LGUs, including wastewater treatment facilities, to address discharging of polluted wastewater into the lake and tributary rivers.

National Greening Program (NGP)
The LLDA contributes to the NGP objectives, which include planting 1.5 billion seedlings in 1.5 million hectares of public lands nationwide, from 2011 to 2016. The NGP also aimed to improve water quality in rivers and to irrigate farmland, reduce flooding, absorb carbon dioxide from the atmosphere, and support the expansion of a wood-product economy. Using its own resources, the LLDA produces tree seedlings for distribution to LGUs and other beneficiaries for planting in their respective areas. Furthermore, the authority undertakes reforestation activities in denuded/designated areas in critical sub-watersheds.

River Rehabilitation Program (RRP)
The River Rehabilitation Program (RRP) takes into consideration the effects of water quality and quantity to the ecosystem. Under the program, the LLDA has organized River Basin Councils all over the region. The "Environmental Army," an aggrupation of community volunteers, has been established and mobilized to help the LLDA in various river clean-up operations. The LLDA has transformed the River Basin Councils and the Environmental Army into empowered partners in environmental governance throughout the lake region.

Shoreland Management Program
The Shoreland Management Program is designed to control pollution and nuisances through the elimination of incompatible elements and uses of shoreland through the LLDA Board Resolution No. 23, Series of 1996, which defines and regulates the use and/or occupancy of the Laguna de Bay Shoreland Areas.

4.2 *Local Government Unit (LGU): Santa Rosa City[4] and a Study in the Santa Rosa Sub-Watershed*

Local government units (LGUs) also play important roles in the Laguna Lake watershed/resource managements. Santa Rosa City takes the head in the lakeshore area and established its own City of Santa Rosa Environmental Code in 2012. The city

[4] Refer to http://santarosacity.gov.ph/

also developed City Planning 2020, and it focuses on harmonizing development and environmental conservation including Laguna Lake watershed/resource management.

Due to the rapid development since 1980s, land use allocation in the Santa Rosa City had dramatically changed from wildlife, agroforestry, sugarcane field, and paddy farm to highways, factories, residential, and commercial areas. Santa Rosa City is a typical case that it faced the environmental degradation and is struggling for better water governance.

In order to evaluate local and nonlocal food consumption behavior and food-health risk in the watersheds, Research Institute for Humanity and Nature[5] (RIHN) conducted a household survey and health examinations in the Santa Rosa Sub-Watershed, one of 24 sub-watersheds that are part of the Laguna de Bay Watersheds. Santa Rosa Sub-Watershed is located 30–40 km from Metropolitan Manila and had a population of about 622,000 as of 2007. The land use changes and population explosion have degraded surface and groundwater in the sub-watershed as well as streams that flow into Laguna de Bay.

Out of 359 households in the Santa Rosa Sub-Watershed, 93 households (26%) eat tilapia from the Laguna Lake on a regular basis, and 257 households (72%) do not. Out of the 93 households who regularly eat local tilapia, 35% of them had household members who did not pass the health check, while 16% of the 257 households had household members who did not. Employing Pearson's chi-square test, the study verified the causality of consumption of freshwater fish from Laguna de Bay and health condition. Results indicated certain linkage of the local fish consumption and health risk.

4.3 Barangay

In the Philippine society, *barangay*, corresponding to village or local community unite, also plays roles as minimum units of social decision-making and collective action. Goto (2004) classified the *barangay* function into two: formal and informal functions. As social capital, barangay's function or its local-level social system contributes to sustainable resource use and environmental management, such as promotion of conservation farming (Cramb 2004) and forest reserve program (Dahal and Adhikari 2008). *Barangays* also take an important role in the Laguna Lake watershed/resource managements. One of the community-based activities or collective action is *Yaman ng Lawa.*

[5] Refer to http://www.chikyu.ac.jp/rihn_e/

5 Community Activities: *Yaman ng Lawa*

The *Yaman ng Lawa* (YNL; Blessings of the Lake, *in Tagalog*) is a social action research for making traditional knowledge science-based. Three projects were conducted as follows:

- Agricultural and domestic waste management project in Santa Rosa City and municipality of Silang (see Ranola et al., 2011)
- Treating excreta and public health project in Santa Rosa City (see Paterno et al., 2014)
- Water resource and fishery management in the city of Calamba, named "Yankaw Fish Garden Sanctuary Project"

This chapter focuses on the third project (see Buen, C.C., 2016).

The YNL Yankaw Fish Garden Sanctuary Project started in 2013 as a participatory grassroots action research conducted by LakeHEAD,[6] 10 fishing *barangays* (villages), local government office of Calamba City, and the Laguna Lake Development Authority (LLDA). The Yankaw Fish Garden Sanctuary established 24 individual *bubo* (habitats), each serving as artificial reef made from branches of a leguminous tree locally known as *camachile* (*Pithecellobium dulce*). Researchers report that *camachile* contains tannin which calms some fish and can induce spawning. The project has laid the basis for sustainability of the Yankaw Fish Garden Sanctuary by developing community-based tools for restoring lake habitat to improve fish populations and harvest as stop-gap measure to restore fishing livelihood and income losses. Program follows the common resource management idea promoted by E. Ostrom (1990). The YNL Yankaw Fish Garden Sanctuary Project organized a lake patrol team and system to watch the fish garden as well as their open fishing ground. Even after the LakeHEAD research project expired in 2015, the fish garden sanctuary has been maintained, and the city of Calamba continues to support their activities with a budget. The LLDA is promoting this type of YNL activity to other lakeshore barangays.

Community Mobilization

Prior to the discussion on the details of the YNL project, a series of consultations and informal workshops were carried out together with the Calamba City[7] Agricultural Services Office, starting in October 2012. The project, designed to become a social action program of the LakeHEAD, has focused on community participation and leadership in communicating the results and outcomes of the project. Community ownership of final information and recommendations of the project is the core strategy to ensure sustainability of the project. A community-based approach was employed and a transdisciplinary (TD) team (fisherfolks, local and national government, and researchers) was formed to conduct the YNL Fish Garden

[6] **Lake h**ealth, **e**nvironment, **a**griculture, and bio**d**iversity. LakeHEAD was the research consortium (RIHN, UP Manila, UP Los Baños, and LLDA) who triggered the YNL activities.

[7] Refer to http://www.calambacity.gov.ph/

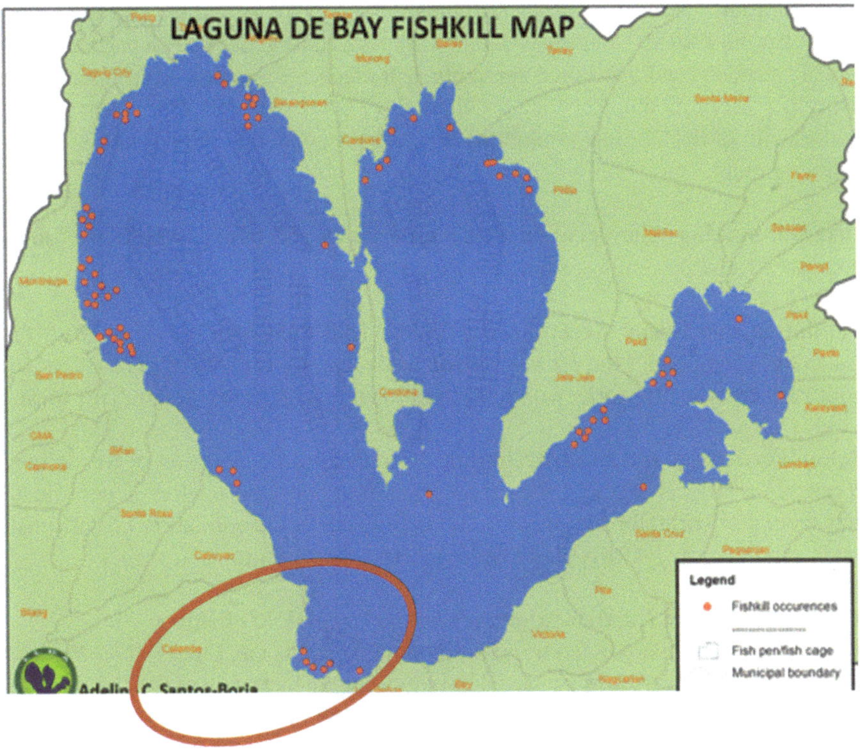

Fig. 7.2 Location of fish kill occurrences, fish pen/fish cage, and city of Calamba (Source: LLDA)

Sanctuary Project. The presidents of the Fisheries and Agricultural Resource Management Council (FARMC) were targeted participants in the process of mobilizing the fishing communities to obtain wider views on issues of leadership and governance.

The TD team first paid attention to fish kill phenomenon (see Fig. 7.2), and the community-based validation research followed five processes:

Consultation They held series of informal meetings to discuss priority problems of the lake and effects on their livelihood, food, and health. The major outputs of this consultation are (i) identification of their availability for participation in the project and (ii) the collective contribution of risk identification through bio-signals[8] and the translation of these information into a map of bio-signals in Laguna de Bay. It indicated that the ecological bio-signals increase in March, April, and May (3 months) while fish kills seem to occur.

[8] Eight types of bio-signals were identified: (1) bad water quality (blackened polluted water), (2) algal bloom, (3) fish kill, (4) loss of lake water plants, (5) reduction in fish catch, (6) presence of chemicals in water, (7) floating of shrimps and *kanduli* (catfish) and presence of white crane, and (8) scum tastes in fishes.

Coordination The process of identifying roles and responsibilities. The major output is that the FARMC and the Calamba City Agricultural Service Department collected, recoded, and submitted the daily fish catch information from open-water fishermen every month. The data provided the TD team and FARMC clear picture of the status of the lake fish availability and presence of community of fishes in their fishing ground.

Decision-making They confirmed the agreements and terms of participation in processes 1 and 2.

Validation and Social Acceptance FARMC held their own dialogues with the respective members in the *barangays* for feedback and updates of their offers for voluntary community works with their traditional "*Bayanihan* (cooperative) spirit."

Policy Formation and Mainstreaming Referring to the elevation of the defined cooperation to local officials, the City Agricultural Service Department coordinate and facilitate works for the project, especially monthly compilation of daily fish catch submitted by FARMC and regular meetings/workshop.

YNL Yankaw Fish Garden Sanctuary as the LakeHEAD Project
Following the TD team research and discussion, the Yaman ng Lawa (YNL) Yankaw Fish Garden Sanctuary was proposed to reduce the impact of fish kills and mitigate the declining fish catch from open water fish ground (see Figs. 7.3). The layout of Yankaw Fish Garden included cluster of 22 Yankaw sites that spread out to a total area of 12 hectares. The Yankaw design is derived from the series of consultations with farmer leaders and open-water fisherfolks, who stated that the 22 Yankaws would not hamper the daily travel and fishing activities for local open fishermen.

In order to manage the fish garden, the following four items were agreed:

1. Basic objectives for the Yankaw are for restoration of local fishes to improve healthy fish stocks in the open waters. Farmer leaders and local community will help ensure no fishing activities in the garden to get reliable fish yield data.
2. The Yankaw shall have buffer rows of Livelihood Yankaw (10) and inner rows of Conservation Yankaw (12). Fish harvesting from the Livelihood Yankaw area will be allowed, and all earnings from the sale of fish catch shall form as funds for improvement, maintenance, and establishment of new sets of Yankaw as agreed upon by the Barangay FARMC. The inner rows of Yankaw are for fish conservation, restoration, and protection to ensure that juvenile fishes grow.
3. A system of schedules of periods of harvesting will be agreed to allow full growth of juvenile fishes and to sustain breeding and restoration of fishes in the open waters.
4. FARMC will continuously conduct information campaign on the role and importance of the Yankaw to the livelihood of the open waters fishermen and to the improvement of stocks of variety and healthy local fish stock in the lake.

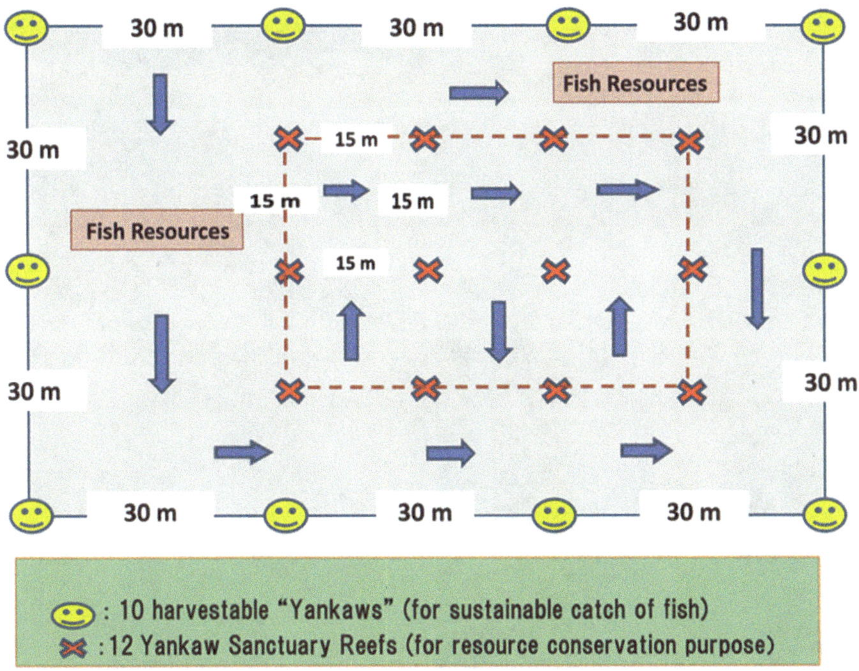

Fig. 7.3 Design of "Yankaw Sanctuary" System for Recovery of Fishery Recourses (Source: Concepcion et al. 2013. Page 281)

Turning Tragedies Into Opportunities

The YNL Yankaw Fish Garden Sanctuary resulted in increase in fish catch as well as fish species (Concepcion et al. 2013 and Interview 2016.). After the LakeHEAD Project (3-year full phase) expired in March 2014, Typhoon Rammasun (Typhoon Glenda in the Philippines) had destroyed the Yankaw Sanctuary Headquarter Tower in July 2014. However, the Calamba fishermen from ten barangays and FARMC have kept maintaining the Yankaw Fish Garden Sanctuary as well as patrolling the fish garden and open-water fish ground. The city of Calamba allocated the budget to their activities and the headquarter tower was restricted in 2016 (see Fig. 7.4).

According to the interview survey conducted in November 2016 (see Fig. 7.5), the fisherfolk leaders share strong motivation to maintain the YNL Yankaw Fish Garden Sanctuary and to continue their activities including lake patrol. Their motivations for participating in the project are based on their understandings:

(i) They had experienced the water environmental degradation and fish species/numbers decrease, either observation or catch, through their daily work.
(ii) They confirmed that the traditional and adaptable Yankaw Fish Garden Sanctuary worked; they found the signs of fish species/numbers restoration.
(iii) When they had witnessed illegal fishing on the lake, individual fisherfolk had nothing that can be done though he was very bitter about that.

Fig. 7.4 Yankaw Fish Garden Sanctuary headquarter and lake patrol schedule table (Source: Photos taken in Nov. 2016)

(iv) The lake patrol activity functioned as lake watching and report system with less additional cost by work shift. The activity also could reduce their stress.

The fisherfolks realized that the YNL collective action linked with their livelihood as well as contributed to the lake governance.

City government of Calamba took the community-based actions into their fishery and water resource management policy with budget. LLDA also focuses on the Calamba case, and community development officers examine how to introduce and promote this policy framework to other fishery communities.

6 Discussion and Conclusion

The national through local level activities for the Laguna Lake watershed/resource management are remarkable cases to examine and discuss Torfing's Interactive Governance. We also need to recall and review Community Mechanism by

Fig. 7.5 Focus group interview and discussion (Source: Photo taken at the Agricultural Service Department, Calamba City, in November 2016)

Hayami (2009) and Otsuka and Kalirajan (2011) as well as rural people's self-organizing activities (Shigetomi and Okamoto, 2014). Many stakeholders, including government and nongovernment and local through international, are involved in the Yaman ng Lawa activities (see Tables 7.2 and 7.3). Hayama (2014) clarified the organizational capabilities of rural Philippine society through the case study of Community-Based Forest Management project. Islam et al. (2011) examine the impact of Community-Based Fisheries Management (CBFM) on fishing households' welfare in Bangladesh and found that fishers in CBFM project areas have improved their access to different assets including social, human, physical, financial, and natural capitals. Marschke and Sinclair (2009) also deal with the case of participatory resource management in Cambodian fishing villages. Nasuchon and Charles (2010) examined the involvement of coastal communities in fisheries management among the countries of Gulf of Thailand – Malaysia, Vietnam, Cambodia, and Thailand. Paunlagui et al. (2016) found that the relationship between social capital and eco-governance was positive. Community mechanism as social capital, which includes formal and informal functions in Asia, is a key to examine interactive governance.

Table 7.2 Four loci of water governance in the Philippines

Loci of water governance	Producer of water policy	Scope of the policy	Participation opportunities	Basis of policy	Resolution of oppositions
National government	National government water agency	National	As prescribed by law	Statues	Mainly through prosecution and litigation
National nongovernment	National nongovernment water institutions	National	As allowed by social arrangements	National demand and stakes on water	Negotiations and litigations
Local government	Local government units	Local	As prescribed by law	Interests and influence of local users	Negotiations and litigations
Local nongovernment	Local water users	Local	As allowed by social arrangements	Local dependence on and use of water	Mainly through negotiations

Source: Malayang 2004, Page 78

Table 7.3 YNL Yankaw Fish Garden Sanctuary Stakeholders

Loci of water governance	Stakeholders
Local nongovernment	**Fisherfolks, fisher households**
	Fisheries and Agricultural Resource Management Council (FARMC)
	Lakeshore barangays (10) and their captains
Local government	**Lakeshore barangays (10) and their captains**
	City of Calamba, Agricultural Services Department
National non-government	Union Galvasteel Corporation (manufacture) as partner
	SM (national retailor) as partner
	History Makers (youth organization) as partner
	Rotary International Philippines (NGO) as partner
National government	**LLDA Resource Management and Development Department**
	LLDA Community Development Officers
	University of the Philippines (UP) Los Baños (Agriculture, Fishery, and Environmental Science)
	University of the Philippines (UP) Manila (Medical Science and Public Health)
International	**Research Institute for Humanity and Nature (RIHN), Japan**

Source: Arranged by author

Appendix

Laguna de Bay, or Laguna Lake, is the largest lake in the Philippines and the third largest lake in Southeast Asia, regarding lake surface area. Table A1 shows some characteristics of Laguna de Bay.

Table A1 Outline of Laguna de Bay

Item	Figure	Note
Geographical:		
Coodinates	14.3935N, 121.1939E	
Lake Surface Area	900 km^2	The largest in the Philippines.
Watershed Area	2,980 km^2	
Average Depth	2.5 m	
Deepest Point	20 m	Diablo (Diablo Pass)
Average Volume	2,250,000 m^3	
Retention Time	8 months	
Regional Area	4,057 km^2	Administrative Jurisdiction
Shoreline	285 km	at 10.5 m elevation
Total Shoreland Area	14,000 ha	at 12.5 m elevation
Administrative:		
# of provinces included (even partly) in the LL watershed	6	Rizal, Laguna, Cavite, Batangas, Quezon & Metro Manila
# of Shoreland Municipalities	35	Laguna-20, Rizal-10, NCR-5
# of Shoreland Barangays	217	at 12.5 m elevation
Region Population	16,248,298	NSO 2015
Lakeshore pop	4,835,834	NSO 2015
Non-Lakeshore pop	11,412,464	NSO 2015
Economic Use & Benefits		
Fisheries Production	150,957 metric tons	in 2012
Aquaculture	63,490 metric tons	in 2012
Municipal Fisheries	87,467 metric tons	in 2012
Business Establishments	About 500,000	estimated, large, medium & small
Electricity	758 megawatts	3 plants, as of 30 June 2015
Industrial Cooling	2.04 billion m^3	lake water yearly
Agriculture	32,684 ha	potential est. total irrigable area
Recreation	Fishing, boating, sailing, swimming	
Floodwater Reservoir, Water Supply (domestic)		

Note: Its biological resources include fish, mollusks, crustaceans, plankton, and macrophysics
Source: LLDA website, arranged by author

References

Buen, C.C. 2016, March 24. *Yaman ng Lawa: Adaptive Co-Management of Lake Resources: The Experience of Calamba City, Philippines*. Presentation at the Special Seminar of Food-Risk CR Project, RIHN, Kyoto, Japan.

Concepcion, R.N. et al. 2013. Yaman ng Lawa Social Action Agenda: The Yankaw Fish Garden Sanctuary. In *Research Institute for Humanity & Nature. RIHN-LakeHEAD Community Forum 2013: Proceedings*, 272–293. ISBN: 978-4-902325-99-7.

Cramb, R.A. 2004, July. *The Role of Social Capital in the Promotion of Conservation Farming: The Case of Landcare in the Southern Philippines*. In 13th International Soil Conservation Organization (ISCO) Conference, Brisbane.

Dahal, G.R., and K.P. Adhikari. 2008 May. *Bringing, Linking, and Bonding Social Capital in Collective Action*. CAPRi Working Paper No.79, 23 p.

Edelenbos, J., N. van Schie, and L. Gerrits. 2010. Organizing Interfaces Between Government Institutions and Interactive Governance. *Policy Sciences* 43: 73–94.

Goto, M. 2004 February. The Functions of Village Organization "Barangay" and Local Society of the Philippines: A Case Study of One Village of Laguna. *Forum of International Development Study* 25: 61–80. In Japanese.

Hall, R., J.C. Lizada, A.C. Rola, C.L. Abansi, M.H.F. Dayo, and M. David. 2014, March 1. *To the Last Drop: The Political Economy of Philippine Water Policy*. Presented at the Philippines Studies Conference, CSEAS, Kyoto University, Japan.

Hayama, A. 2014. Forms of Collective Actions in a Dyadically Woven Local Society: Case Study of Rural Philippines. In *Local Societies and Rural Development: Self-organization and Participatory Development in Asia*, ed. S. Shigetomi and I. Okamoto, 129–162. Cheltenham: Edward Elgar.

Hayami, Y. 2009. Social Capital, Human Capital and the Community Mechanism: Toward a Conceptual Framework for Economists. *The Journal of Development Studies* 45 (1): 96–123.

Islam, G.M.N., T.S. Yew, N.M.R. Abdullah, and K.K. Viswanathan. 2011. Social Capital, Community Based Management, and Fishers' Livelihood in Bangladesh. *Ocean & Coastal Management* 54: 173–180.

Laguna Lake Development Authority. Official site: http://www.llda.gov.ph/

Malayang, III, B. 2004. A Model of Water Governance in the Philippines. In *Winning the Water War: Watersheds, Water Policies and Water Institutions*, ed. Rola, A.C., Francisco, H., and Liguton, J.P.T., 59–83. Philippine Institute of Development Studies and Philippine Council for Agriculture, Forestry and Natural Resources Research and Development; Makati City, Philippines.

Marschke, M., and A.J. Sinclair. 2009. Learning for Sustainability: Participatory Resource Management in Cambodian Fishing Villages. *Journal of Environmental Management* 90: 206–216.

Nasuchon, N., and A. Charles. 2010. Community Involvement in Fisheries Management: Experiences in the Gulf of Thailand Countries. *Marine Policy* 34: 163–169.

Ostrom, E. 1990. *Governing the Commons: The Evolution of Institutions for Collective Action*. Cambridge: Cambridge University Press ISBN 0-521-40599-8.

Otsuka, K., and K. Kalirajan, eds. 2011. *Community, Market and State in Development*. New York: Palgrave Macmillan ISBN: 978-0-230-27458-7.

Paterno, R.P., J.Z. Galvez-Tan, L.A. Palapar, A.N. Bermudez, and A.J.B. Guiang. 2014. *The Yaman ng Lawa (Blessings of the Lake) Initiative in Santa Rosa City, Laguna, Philippines: An inclusive, Participatory Approach to Publicpolicy Development*. Presentation Paper, 15th World Lake Conference, Perugia, Italy.

Paunlagui, M.M., M.R. Nguyen, and A.C. Rola. 2016. *Social Capital, Eco-Governance, and Natural Resource Management: A Case Study in Bukidnon, Philippines*. Working Paper No. 03-04, Institute of Strategic Planning and Policy Studies, UP Los Banos. 27 p.

Ranola, R.F., F.M. Ranola, M.C.S. Casin, M.F.O. Tan, J.T. Dizon, and M.N.Q. Herrera. 2011. *Assessment of Domestic, Fishery and Agricultural Wastes in Silang-Santa Rosa Sub-Watershed: Survey Results for the Yaman ng Lawa (YNL) on Adaptive Community Waste Management (ACWM) Project Sites*. LakeHEAD Annual Report No.1 FY2010. RIHN, Kyoto, Japan.

Research Institute for Humanity and Nature (RIHN). 2011 July. *Managing Environmental Risks to Food and Health Security in Southeast Asian Watersheds*. Progress Report No. 1. FY 2010–11. ISBN 978-4-902325-68-3.

———. 2013a. RIHN-LakeHEAD Community Forum 2013. Yaman ng Lawa: Adaptive Co-Management for Sustainable Use of the Wealth of Laguna Lake. November 7–8, 2013. BP International Makiling, Los Banos, Laguna. Proceedings. ISBN 978-4-902325-99-7.

———. 2013b. *Proceedings: Community Forum 2012. September 27–28, 2012*. Development Academy of the Philippines, Tagaytay City. ISBN 978-4-902325-83-6.

———. 2014 March. *Managing Environmental Risks to Food and Health Security in Southeast Asian Watersheds*. Final Year Report FY 2013-14. ISBN 978-4-906888-03-0.

Rola, A.C., C.L. Abansi, R. Arcala-Hall, J.C. Lizada, I.M.L. Siason, and E.K. Araral Jr. 2015. Drivers of Water Governance Reforms in the Philippines. *International Journal of Water Resources Development* 32 (1): 135–152.

Rola, A.C., C.L. Abansib, R. Arcala-Hall, and J.C. Lizadad. 2016. Characterizing Local Water Governance Structure in the Philippines: Results of the Water Managers' 2013 Survey. *Water International* 41 (2): 231–250.

Shigetomi, S., and I. Okamoto. 2014. Local Societies and Rural People's Self-Organized Activities: An Analytical Framework. In *Local Societies and Rural Development: Self-organization and Participatory Development in Asia*, ed. S. Shigetomi and I. Okamoto, 1–19. Cheltenham: Edward Elgar.

Torfing, J., B.G. Peters, J. Pierre, and E. Sorensen. 2012. *Interactive Governance: Advancing the Paradigm*. Oxford: Oxford University Press 265 p.

Tadayoshi Masuda is Associate Professor of Agricultural Economics and Management in the Department of Agricultural Science, Faculty of Agriculture, Kindai University, Nara, Japan. His research interests include watershed/farmland/agroforestry management and community mechanism, global value chain of agricultural products (e.g., tea, coffee, soybeans) and regional development, conservation agriculture (CA) technology/knowledge, and policy mix. Prof. Masuda received his PhD in Agricultural and Resource Economics from the University of Hawaii in 2007, MA in Food Research/International Development Policy from Stanford University in 1997, and BA in Agricultural and Forestry Economics from Kyoto University in 1989.

Chapter 8
Action Research Aimed at Contextualizing Partnership in Irrigation Water Management in South Sulawesi, Indonesia

Kazuko Oguni

Abstract This chapter examines the process and outcomes of action research on irrigation management in Sulawesi, Indonesia. The objectives of the research were to provide an opportunity for old and new stakeholders to meet, discuss, and make decisions together and to provide experience of using interactive processes in implementing agreed-upon solutions. By jointly experiencing challenges and successes based on mutual decision-making with moderate support from the researchers, a space for participation and creating contextualized roles and responsibilities was formed based on the sociocultural situation. Moreover, this occurred regardless of whether the participant was on the "governing" or "governed" side. This chapter addresses the dynamic integration of local customs and government-led systems facing mutual discontinuity, which is a concern from the cultural perspective of interactive governance.

Keywords Action research · Indonesia · South Sulawesi · Local context · Participatory irrigation management · Collaborative interaction

1 Introduction

This paper attempted to depict the results of and indications provided by action research conducted in 2014–2015 (as part of a research project spanning from 2011–2015) in order to examine the efficiency of collaborative practices among local stakeholders in irrigation water management in South Sulawesi, Indonesia. The stakeholders consisted of both farmer organizations and multilayered local government employees as well as local and international researchers.

Two main settings determined the local form of governance in this case. The first was the long local history of irrigation water management at the grassroots level since the lordship era. As a result, local residents have a specific common image of how and who should be responsible for daily management. These socially

K. Oguni (✉)
Faculty of International Welfare Development, Nihon Fukushi University, Aichi, Japan
e-mail: oguni@n-fukushi.ac.jp

K. Otsuka (ed.), *Interactive Approaches to Water Governance in Asia*,
https://doi.org/10.1007/978-981-13-2399-7_8

constructed images thereby affect attitudes toward and acceptance of new services and institutions introduced by the government. The second consists of national economic and political changes that have taken place over the past decade in the Republic of Indonesia. The spread of decentralization has encouraged political initiative at a more local level, for example, through regency, county, and village governments. This appears to be one of the factors supporting interactive governance between the public sector and farmers' organizations. It has also provided increased opportunities for villagers to be part of the village and the county administration. Village elites who have double self-identities as one "who governs" and one "who is governed" play an essential role in connecting the discontinuities between government policies and daily farming practices as well as between government staff and farmers and between "traditional" and "modern" systems.

Action research was therefore planned to create practical opportunities for new and old stakeholders of irrigation water distribution to familiarize themselves with each other's roles. Discontinuity between traditional water managers (so-called *Mandoro Jene* (hereinafter referred to as MJ)) and government-led modern farmers' organizations (so-called P3A[1] in the Indonesian language) was observed during the initial period of research in 2011–2012. Integration of old and new participants was subsequently required to address the mutual distrust, by realizing collaborative practices and sharing experiences of success.

This article focuses on the implementation of action research in South Sulawesi, Indonesia, from 2014 to 2015. The post-research situation in the field in 2016 is also discussed. Important methodological considerations aimed at stimulating, strengthening, and stabilizing collaborative interaction between farmers, and delegates of farmers' organizations and public workers, all of whom are in a position to govern water distribution in the area, are also introduced. The involvement of researchers as mediators in the action research was also a target of observation.

The observation in this chapter focuses on the way in which people contextualized collaborative behavior based on their experiences, including the sharing of local knowledge, during the action research period. Knowledge is constructive and "the outcome of many decisions and selective incorporations of previous ideas, beliefs and values" (Long and Long 1992; Long 2001:242–243). Dialogues and negotiation among multiple stakeholders in irrigated water distribution, therefore, were recorded and carefully observed to understand their role in the process of internalizing knowledge in the field.

[1] P3A = *Perkumpulan Petani Pemakai Air*: farmer water user's association (hereinafter referred to as P3A).

2 Background: Transition in Unique Local Irrigation Systems

2.1 *Outline of the Research Project and Specific Concerns in Sulawesi*

The overall title of this project, which was carried out from 2011 to 2015, was "Designing Local Frameworks for Integrated Water Resources Management[2]." The objective was to provide ideal water management at the regional level, as well as a knowledge-based framework and specific practices to help its realization (with particular focus on joint management by water users, particularly agricultural households, to identify the details of "ideal management" and conditions for and significance of achieving this)[3].

The project took a multidisciplinary approach that included hydrological, cultivation, and climate studies as well as analysis of cultural anthropology. The objective was to propose a knowledge-based structural framework and functions of water resource management to local-level stakeholders who play an essential role in implementing integrated water resources management (IWRM). The project therefore involved considerable exchange of scientific evidence on particular water cycles with the wide range of stakeholders involved in water management and use. The main goal was to develop cooperation between science and society in order to stimulate the co-creation of desirable local water resource management.

Individual area teams were in charge of independent field research. At the initial stage, the co-researcher of the Sulawesi team, who also runs a local NGO, wanted to know whether it would be possible to incorporate MJs as technicians within the P3A organization formed by the government. Based on this enquiry, the historical significance of the MJs was clarified, establishing whether or not they can work within the P3A organization and, if so, considering what role they might play. As a first step, the historical role of MJs as water managers was examined in order to both understand their social significance and consider the potential for collaboration in future organizational irrigation management.

One important issue of organizational management in this case was the immaturity of the P3A federations that were later institutionalized by the government. The 11 federations were organized to coordinate all 130 P3As in the area. However, their roles and responsibilities were not effectively conveyed to local farmers, even by federation leaders themselves. This was in contrast to the P3As, which were gradually accepted as a new form of organizational irrigation management for 20 years.

---

[2] http://www.chikyu.ac.jp/rihn_e/project/C-09.html

[3] Target areas included Turkey, Japan, and Indonesia. At first, five research targets were identified; however, after the initial 2 years, focus gradually shifted to the three main sites where prior research had been implemented.

Two survey objectives were therefore decided, the first being to depict an overall image of local systems, especially MJs and other government-organized "modern" stakeholders such as P3As and P3A federations. The responsibilities of each in irrigation water management in the field were also confirmed. Based on the results of this first survey, the second objective was to determine a practical and methodological approach to stimulate interaction aimed at improving collaborative irrigation management. Accordingly, any changes that occurred as a result of the mutually interactive processes were recorded. As observation focal points, these two issues, the integration of MJs at the grassroots level and strengthening of P3A federations at the mezzo level, were targeted.

2.2 Implementation Structure

A team of researchers from Japan and Indonesia was organized (Fig. 8.1) to implement intense surveys, interviews, and meetings during short-term visits. "Local researchers" (local NGO staff) also visited the target villages on a day-to-day basis, collecting data periodically.

Such remote management could only be realized because of the experiences and respect shown by the researchers toward the local people during their time as NGOs in the past decade. The co-researcher of the Sulawesi team, a professor at the National University of Sulawesi, also runs the above local NGO. This NGO has been in operation for more than 10 years in an area receiving water from the Bili-Bili Dam, supporting female empowerment, and working with the local P3A. Experienced NGO field staff were therefore employed as "local researchers." Their existing networks and local knowledge resulting from previous experience as facilitators was essential in designing action research aimed at the realizing and prioritizing the roles of local farmers. The NGO staff regularly consulted with the

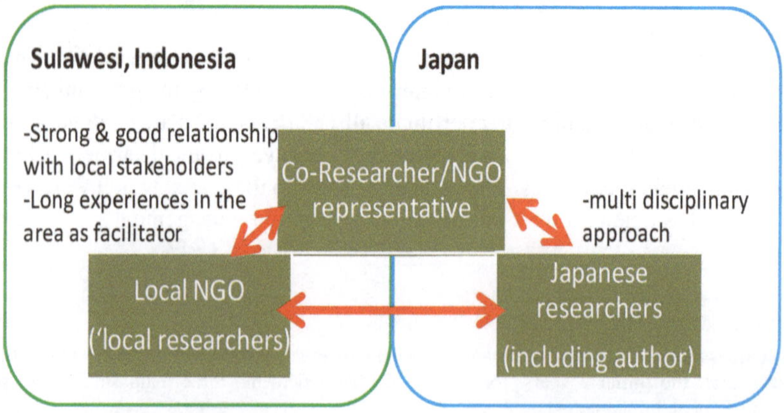

Fig. 8.1 Implementation structure (Source: Compiled by the author)

co-researcher, who continuously traveled between Sulawesi and the research institute in Japan. This mobile structure helped the Japanese researchers keep up with the dynamic changes occurring in the field throughout the action research.

2.3 A History of Irrigation Development and Current Conditions in the Focus Area

The Sulawesi research field was located in the Bili-Bili irrigation area, Gowa Regency, South Sulawesi (Fig. 8.2). Within the beneficiary area, which is divided into three irrigation systems, we focused on Kampili region (10,545 ha), the oldest of the three systems.

The Gowa Regency, one of the three major regencies in South Sulawesi, was founded around the twelfth century. The area has retained characteristics of the lordship system still in existence (Assegaf 2008; Kulle and Tika 2008a,b). The Dutch ruled the area in 1669, during which time the Kampili weir and main water route were constructed. The current province of South Sulawesi was founded in 1964, after establishment of the Indonesian state in 1945 and dissolution of the State of East Indonesia in 1950.

In 1978, a yen loan was provided for the integrated development of the Je'ne Berang River (total loan capital, 5.5 billion yen). The multipurpose Bili-Bili Dam construction was completed in 1999. Prior to this, a water user's association, the so-called P3A, was established in 1996, and several training sessions were provided. Since then, the local NGO (2001–) and Japanese government (2004–2007) have continued to implement projects aimed at strengthening the P3A.

Figure 8.3 highlights the structure of government service providers and farmer organizations involved in irrigation water management in the area. Several changes have occurred in recent years, including the transfer of direct governance of P3As and federations from the office of water resource management to that of agriculture in 2014. As a result, confusion has arisen among farmers over who they should contact, and in what way, in order to have their voice heard. This further highlighted the need for collaborative practices aimed at strengthening cooperation between the two offices.

Moreover, at the grassroots level, rural life has also undergone change. The general image of villages in the focus area is that of communities with the local language, Makassarnese, spoken as the first language. The main religion is Islam, as is standard in Indonesia. Giving T village as an average example, the entire area consists of approximately 250 ha (rice fields, 65%; crops, 8%; housing, 5%; and brick-making, 22%), with a population of around 3000 (men, 47%; women, 53%) and around 770 households (Statistical Dept. data, 2010).

The rainy season starts in October and lasts until April, with the dry season running from May till September. Individual farms consist of a relatively small area of land, equivalent to roughly 25 acres. An unstable water supply and increasing prices

Fig. 8.2 The river basin and Bili-Bili irrigation area (Source: "Designing Local Frameworks for Integrated Water Resources Management (2011–2015)" by the Research Institute for Humanity and Nature)

have resulted in many farmers seeking a secondary income. This second source of income is considered more stable than irrigated rice production, for example, brick-making. Nevertheless, the irrigation development mentioned above has a wide range of impacts on the daily life of farmers, for example, adding a rice cultivation cycle during the dry season. Thus, rice cultivation remains a priority not least because it is a staple food. Accordingly, many farmers suggested that they intend to continue growing rice even if they turn a profit from their second income. These intentions again confirm the importance of exploring ways to strengthen the multi-stakeholder governance of irrigation water management.

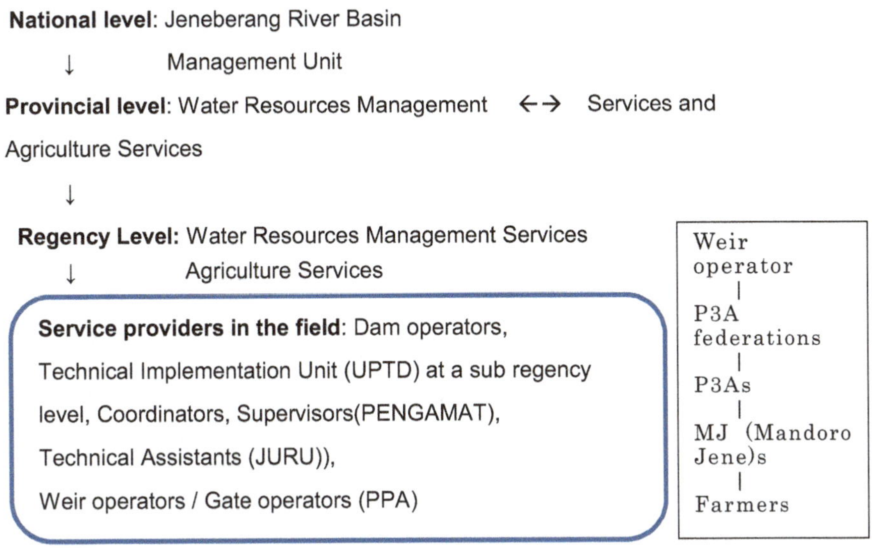

Fig. 8.3 Providers of government and farmer organizations in water management in the Bili-Bili irrigation system (Source: Compiled by the author)

2.4 *Profile of Traditional Mandoro Jene (MJ)*[4]

According to the results of a questionnaire implemented in 2011 by the local NGO, traditional water managers (*Mandoro Jene*, MJ), whose role is to distribute water to paddy fields in the Bili-Bili irrigation system, are an average of 45 years old and work across an average area of 25 ha. However, the actual role of an MJ is rather trivial, involving manual labor related to day-to-day water distribution. Once a gate has been opened, the MJ monitors the flow of water as well as communicates requests from farmers to local government staff. Thus, the "management of water routes," in practical terms, involves daily patrols and the cleaning up of weirs. The MJ works alone, day and night, essentially committed to cleaning up.

Remuneration of MJs takes the form of "*Ase Mandoro*" (Ase = rice plant), which is collected directly from rice fields at harvest. The MJ receives 1–3 *basse* (buckets, roughly 5 liters) of unhulled rice as payment. If the MJ arrives late for collection of payment, no payment is received. At present, according to the rules of the P3A, remuneration should be collected by the P3A and then divided between the MJs. Alternatively, the MJ should collect the payment and share it with the P3A.

Villagers describe the MJ as both "*dihormati*" (respected) and "*ditakuti*" (feared) because of their "*jujur, adil, rajin*" (honesty, fairness, diligence). These characteristics have become a *jaminan* (guarantee) of the MJ, and they are often

[4]All farmers interviewed suggested that only males can work as an MJ since some tasks are considered dangerous, such as patrolling at night and cleaning the gate. Nevertheless, two female MJs were employed in Kampili in 2015, although they were referred to as "exceptions" because of their economic condition and knowledge of MJ work.

repeated by those currently employed as an MJ. Historically (during Dutch rule), the conditions for selection of an MJ were as follows: no farmland, no form of employment, and no secondary income. Such was the case that one rich farmer confessed, "I wouldn't want to be an MJ. It's a dirty, tough job. I wouldn't want my child to be an MJ either."

Emerging from the initial survey in 2011–2012, it was therefore surprising to learn that the MJ works alone. A large quantity of rubbish is collected by the MJ during cleanup. In Indonesia in general, there is a tradition of mutual support, "*gotong-royong*," and in this area also, the village mayor and leader of the sub-village act to mobilize residents twice a year to clean the water routes (prior to opening of the gates). In contrast, the day-to-day cleanup of the water routes is considered the work of the MJ only and sometimes involves removing dead animals such as dogs. This is therefore a very difficult role for MJs, who tend to be Muslim. Nevertheless, MJs carry out their work diligently and without complaint. The MJ has no role in important decision-making such as when to open or close the gates, having no organizational responsibilities. Rather, the MJ works alone in quiet accordance with the desires of local farmers. Patron–client leadership, affected by the lordship system, has been inherited by government leaders who tend to be descendants of noble blood. This charismatic way of governing villagers perpetuates the isolated role of MJs under direct paternalistic management by the government.

2.5 Social/Economic Changes and Loss of Sympathy Among Stakeholders

As mentioned above, under the lordship system, MJs were subjected to strong leadership and clear economic disparity. Initially, their lack of land and work was significant in terms of helping people out of poverty. The personal characteristics of "honesty, diligence, and fairness" were subsequently accepted and internalized as being important in actual water management, leading to sympathy and an understanding among individuals. In turn, MJs gained respect and were perpetuated as part of the local system, building pride in their work as well as local rationale.

However, with changing times and the diversification of livelihoods, this sympathy has changed. Unlike the lordship system, society is now less hierarchical, and there is more choice in terms of work. Furthermore, following the formation of P3A organizations by the government, overall empathy toward MJs decreased. MJs now often feel they are unfairly treated, leading to an increased sense of burden. Thus, confusion has arisen among related parties in terms of the role of MJs within a contemporary context, further emphasizing the need for discussion on how traditional

MJs and modern P3As can work together. A new understanding is required, one more suited to the current age. This became a major factor in our action research planning.

3 Action Research to Create a New Channel for Participation

3.1 "Action Research" in the Case of Sulawesi

Yamori (2010) describes action research as "joint social implementation by researchers and research subjects, who share their thoughts on how society should be." Situations requiring action research include cases that require an "adjustment of values" (Yamori 2010). In this case, the following required adjustment: awareness among those responsible for water management and practical consideration of individual skills of MJs as well as repositioning of MJs within current organizational water management, in line with the diversification of rural villages.

However, action research as an academic method does not provide certain assessment. As pointed out by Edelenbos et al. (2010), "The reliability and validity of action research can be measured by seeing to what extent action, based on research, really resolves problems." The concrete purpose, process of implementation, and various statements of those involved in the action research will be described in the following section, to determine what the project brought to the field.

3.2 Purpose and Methods of the Action Research

The objectives of the action research were to provide opportunities for stakeholders to meet, discuss, and make decisions together and provide experience of working together in implementing agreed solutions. Improved coordination and communication among stakeholders was expected to result from various interactions during the research.

In practice, a series of meetings were held and each meeting recorded. Through these meetings, a simple action plan was then agreed (an opening/closing schedule for the gates, priority given to downriver areas). During the planting period, intensive interviews were conducted mainly by local researchers to clarify communication and networking among stakeholders with regard to daily water management. The changing relationship between MJs and the various stakeholders was also noted.

3.3 *Implementation of Action Research*

3.3.1 Indicator Area/Participation in the Action Research (2014–2015)

In 2014, action research was implemented in Kampili area, the oldest irrigation system in Bili-Bili, thereby becoming the focal point of this research. Three observation targets were indicated as follows. Firstly, 17 P3As were chosen for observations of water coverage during the research period (Fig. 8.4). Secondly, 13 stakeholders from different positions, including 3 P3A federation leaders, 3 gate operators working at the "bottleneck gate" (recognized in initial meetings), and 2 MJs, were selected as subjects for interviews conducted weekly between February and August with the aim of observing how information (regarding gate control, water problems, weir cleaning, meeting schedules, and so on) was delivered. Thirdly, the results were combined for analysis, together with data on water volume at each gate.

3.3.2 Action Research Meetings as a Method of Social Preparation

The aim of the action research was not only to improve collaboration but also to create records and carry out observations that appropriately meet practical and academic methodological requirements. At the implementation stage, several small-scale meetings known as action research meetings (ARM) were held, starting in February at the end of the rainy season in preparation for opening of the sluice gates to allow dry season cultivation in April. Regional meetings were started upstream and then gradually moved midstream and downstream, forming issue-based groups. In response, repairs to broken gates and other small actions were implemented at an early stage in each area. In February/March, common issues were determined and shared, and an "action" plan was proposed and implemented. The plan involved opening and closing of the Kampili main sluice gate as well as gates in two critical areas, followed by cultivation based on the opening/closing schedule.

3.3.3 Interactive Planning of Simple Actions Relating to the Whole of Kampili

Through 14 ARM conducted between February and May 2014 across the entire Kampili area, three locations were identified as bottlenecks for water distribution. ARMs were carried out with support from the local researchers, with P3A federation chiefs acting as conveners as well as a bridge between P3As and village governments. At this stage, researchers carefully observed how the information, meeting schedule, results of discussion and repair plans, etc. were delivered. A plan for opening of the water gates in order to distribute water downstream first was subsequently proposed and a strategy decided.

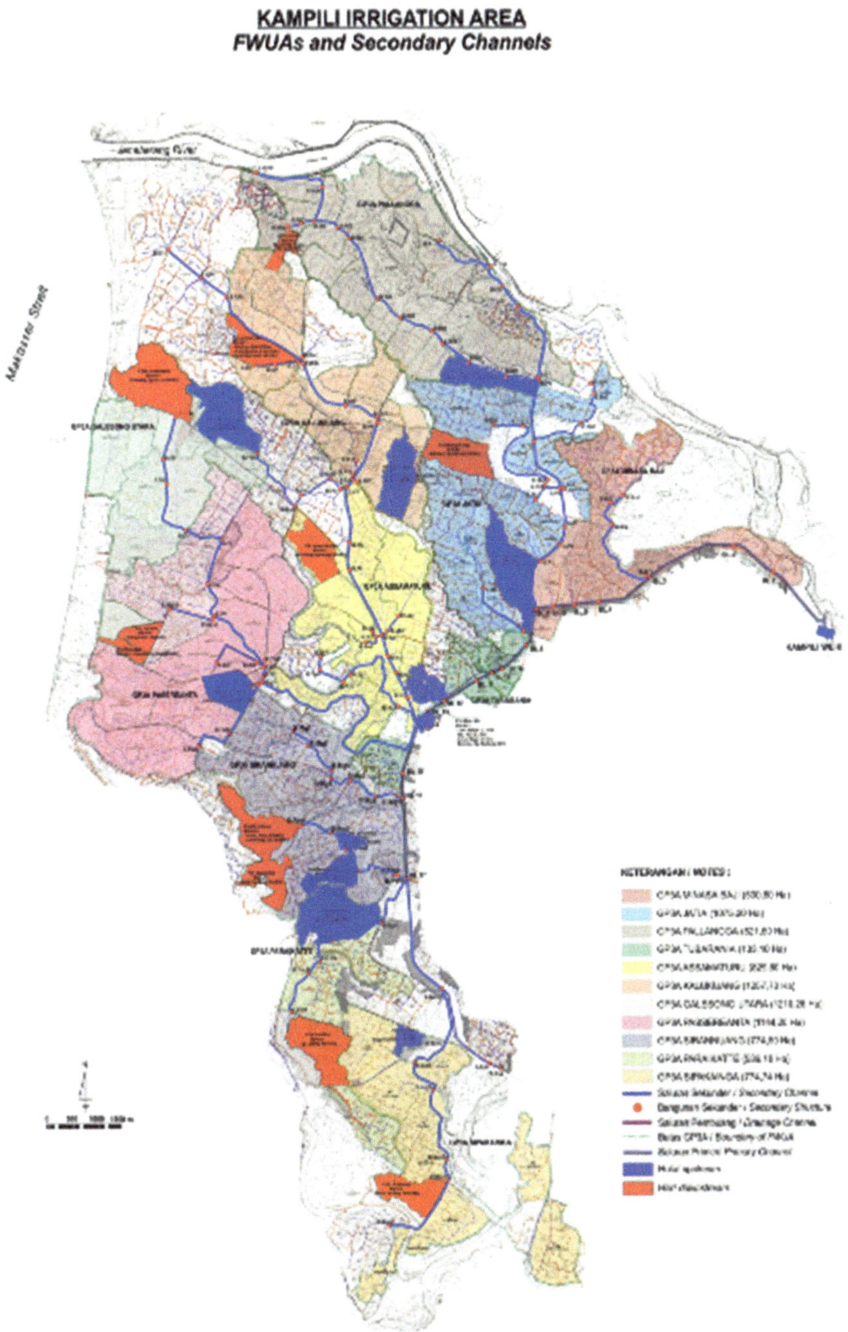

Fig. 8.4 Indicator area (P3As) (blue = upriver, red = downriver of the secondary canal) Source: "Designing Local Frameworks for Integrated Water Resources Management (2011-2015)" by the Research Institute for Humanity and Nature

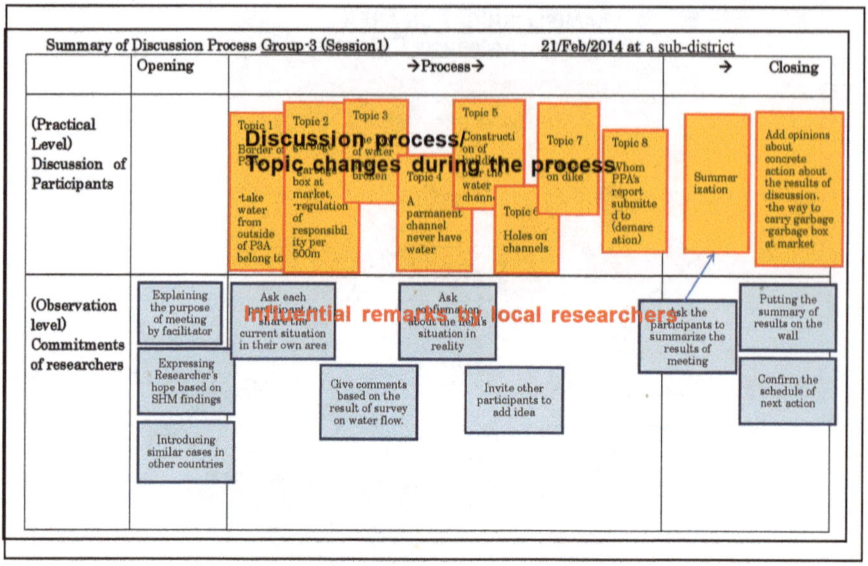

Fig. 8.5 Process flow diagram (Source: Compiled by the author)

3.3.4 Observation Methods During the Action Research

Several tools were prepared for use in the documentation and recording of the action research. Local researchers attended ARMs as often as possible to record the discussion process, making detailed minutes using an IC recorder. Especially in the initial period, the process flow was carefully observed and a flow diagram created (Fig. 8.5 to help clarify key issues, important statements, and necessary changes. The diagram was also used to monitor the local researcher involvement and avoid unnecessary interruptions. This documentation helped the research team understand the context of each discussion even long distance from Japan.

Regular visits by local researchers to the field between February and May were also implemented in order to carry out a communication survey regarding daily water management. These visits often included intensive interviews of MJs regarding the most important periods of rice cultivation (planting out and harvesting) in order to determine the water distribution to terminal water channels. The ongoing concerns of researchers regarding irrigation water management on-site seemed to help the stakeholders become more conscious of their responsibilities, encouraging the need for collaboration.

Fig. 8.6 On-site feedback meeting (Photo taken by NGO Pelangi, August 2014)

3.3.5 On-Site Feedback of "Under-Processed" Data to Farmers (Fig. 8.6)

During the implementation period, researchers created opportunities for some of the survey results to be made available to local farmers. Data was chosen to fit the concerns of the participants, and visible tools such as maps and photographs used to improve overall understanding.

These opportunities for feedback were also helpful for researchers in terms of examining the relevance of the data, allowing the focus of each survey to fit the actual needs of residents in the field.

At the end of the action research period in March 2016, a seminar was held to provide final feedback from researchers to the P3As/P3A federations. Each document used for information delivery, at autonomous meetings and for gate control, was compiled into a handbook (Fig. 8.7) along with the results of the various discussions. The handbook consisted of three parts: (1) an outline of activities (mainly meetings, action plan decision-making, and water distribution schedules), (2) the role of each stakeholder (resulting from discussions), and (3) a comprehensive appendix, with examples of letters and proposals used by local stakeholders during the action research period. The handbook was created based on the requests of the stakeholders themselves, who wished to continue using the action cycles.

Fig. 8.7 "Localized" handbook (Photo taken by the author, March 2016)

4 Assessing the Impact of the Action Research

The following participant statements show the impact of interaction during the action research.

4.1 Scheduled Water Distribution Through Gate Control

Typical statements made by both farmers and public workers during meetings and interviews at the end of the cultivation period/water distribution during the dry season in 2014–2015 included the following:

> We were afraid that things would go wrong if we did anything differently to normal (e.g., if we grew the crop earlier than usual it might be eaten by mice), but in fact it was fine. We would like to do it (the agreed action plan) the same way next year. (P3A federation leader, August 2015)

> The dry season came earlier this year, but thanks to the opening/closing schedule there were no arguments over water. (Gate operator at one of the "bottleneck gates," August 2015)

The federation leaders and gate operators in particular therefore showed their satisfaction with the action research and were strongly motivated to develop an interactive cycle by continuing to (1) hold meetings and (2) create an action plan for gate control aimed at distributing the water downriver first.

In contrast, stakeholders at the grassroots level such as MJs sometimes failed to understand the role of the research as it was planned and implemented in the field. However, at the same time, they also noted that their own daily water distribution during the season had improved due to "the gate opening schedule and through improved management thanks to the increased availability of the gatekeeper". In this way, the internalized impact of the action research was also determined from statements at the grassroots level.

An unexpected desire for better planning also began to emerge.

> We need to review the *rendeng* (rainy season) schedule first in order to improve *gadu 1* (the first dry season cultivation, only targeted during the action research period). (P3A federation leader, August 2015)

Long-term planning for irrigated water distribution does not seem to have been seriously considered by villagers in South Sulawesi region during the past 20 years. Commitment to creating a future long-term plan based on their own experiences was therefore important.

The author visited Kampili during the dry season in 2016 to confirm the implications of the action research in the field. As a result, the scheduling of gate control was found to have been autonomously implemented, with only minor modifications, during the current dry season. In fact, they had also attempted to implement it during the rainy season in November 2015, without result, due to serious droughts in September to October 2015. One leader of a P3A federation who showed remarkable leadership during and after the action research emphasized "The most definitive thing is that we successfully achieved consensus-building regarding distributing water throughout the whole of the Kampili area, downriver first. We will continue to do this, because we have now experienced it. The downriver people will 'rebel' if we stop."

4.2 Mutual Recognition Has Reduced "Barriers" to Communication

4.2.1 MJ Involvement in P3A Organizational Networking

Several significant changes were observed in terms of rebuilding reliance between traditional MJs and modern P3As. For example, the MJ was previously described as being connected to the P3A; however, after daily communication aimed at efficient water distribution during the two dry seasons (Fig. 8.8), the P3A and MJ are now considered a single unit. However, this explicit change in recognition of collaborative roles is only likely to apply to those directly involved in the action research meetings over the last 2 years. Nevertheless, as explained above, several MJs working at the bottom of the organizational structure also experienced improvements in communication with other participants, such as the gatekeeper, facilitating smoother daily water management.

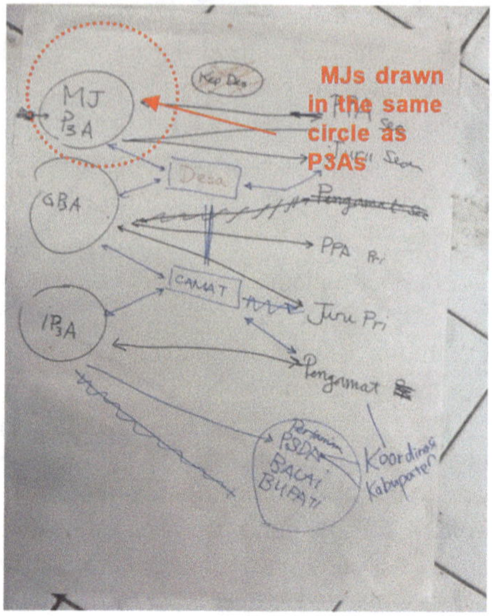

Fig. 8.8 Drawing created from the meetings in Kampili regarding the role of MJs and other stakeholders, August 2015 (Source: Compiled by the author)

4.2.2 New Proposals Aimed at Strengthening Collaboration with the Village Government

One of the most remarkable statements made in relation to organizational capacity development was a proposal for more collaboration with the village government. P3A federation leaders and P3A leaders both mentioned the responsibilities of village authorities at the self-assessment meeting in August 2015:

> Why such low levels of contribution to date?
>
> Water management goes beyond village boundaries (and so coordination by village mayors should be carried out).
>
> If it's not within the village, repair costs are not paid (although the problem directly affects our own area as well).

Accordingly, one village mayor replied: "The P3A only contacts me when there is a problem, and I have never been invited to P3A meetings so the information does not get shared" (August 2015, Kampili).

These dialogues emphasized the need for further discussions among participants as well as the need for P3As to function as part of the village government and the inclusion of village mayors in P3A federation meetings.

During the follow-up visit in 2016, one P3A federation leader who also works as a temporary assistant of the village government noted that documents resulting from the action research, especially those concerning gate control and water distribution schedules, were brought to the annual village development planning meeting and utilized to address issues related to irrigation water management such as gate repairs: "We brought the schedule used during the dry season last year to the irrigation planning committee as an example. Then it became the basis for further plans." Accordingly, active involvement of the village government attracted residents concerned about irrigation water management, raising awareness among villagers.

5 Discussion: Methodological Effectiveness of the Action Research

5.1 Presence of Local Researchers as a Motivational Factor and Communication Bridge

Throughout the action research period, the presence of local researchers was found to motivate cooperation between officials and farmers. Day-to-day visits and questions about daily communication and information delivery seemed to encourage proactivity, encouraging both farmers and officials to collaborate on gate control and water distribution. According to the results of an informal interview with the three local researchers working in the field, their roles in the action research could be summarized as follows:

1. Arrangement of the initial meeting (subsequent meetings arranged by participants themselves)
2. Facilitating discussion during meetings (no manipulation)
3. Assistance with documentation (maps, organizational structure, names and phone numbers, proposals for the regency office, photographs)
4. Acting as a bridge for information flow, for example, between upriver and downriver, farmers and public workers, and MJs and P3As
5. Acting as a bridge between farmers' organizations and the local government, helping to implement proposals related to Gowa Regency
6. Assistance with data collection. In this context, it was important that data was collected not only by researchers but through cooperation with residents to ensure ownership of the collected data
7. Recording of all processes that they are involved in and submission to the project team (from interviews with three staff members, September 2016)

Thus, the roles of the researchers went beyond the image of academic researchers, and they integrated smoothly due to their previous long-term experience as NGO facilitators, thereby encouraging rural participation. Interestingly, this encouragement was often the subject of corrective guidance during the initial meetings in 2014, due to concerns that it might result in overcommitment during objective recording of the qualitative primary data. However, it became apparent that their honest concern of farmer's life also ensured honest data collection.

5.2 Locally Driven Scientific Data Strengthened the Organizational Capacity of Potential Leaders

In line with the role of researchers mentioned in 4-1, a range of data was also essential for ensuring interaction among stakeholders. Simple data on water volume became a trigger in helping farmers understand the issues at hand, thereby encouraging cooperation between public workers and leaders of farmers' organizations. P3A federation leaders and one technician, in particular (a county worker in the field), showed remarkable changes in terms of their roles throughout the action research period. These potential leaders of both farmers and public officials began to look for logical explanations and became highly concerned over the integration of scientific data and daily practice, thereby fulfilling one of main goals of the research project.

As mentioned in Sect. 2, one role of the researchers was to share data as on-site feedback, even though the data was not complete, in order to facilitate an academic conclusion. The data offered here was localized as fast, simple, and concrete, allowing it to be more easily understood. For example, it included data related to improved cultivation schedules and existing water volumes in each water channel and gate, issues repeatedly referred to in meetings.

As a result, the data offered by researchers helped potential leaders be more persuasive when they explain the schedule of water distribution to farmers during the following season in 2016. The knowledge and skills to make presentations based on scientific data seemed to fit with the self-identity of the leaders as a relatively educated elite within the villages, helping increase self-esteem. This then encouraged them to take greater initiative in interacting with other stakeholders to achieve better overall water management.

It was also noted that in 2016 the researcher's perspective had become developed and internalized within the leaders. One P3A federation leader stated that "Our research should go ahead." Thus, accordingly, the simple acts of scheduling water distribution and documentation worked to increase awareness among the various stakeholders, not just farmers' organizations such as the P3A and P3A federation but also village and county officials and related regency offices. These findings suggest that academic data as feedback should be shared with concerned people and contextualized locally during the action process, even if it is yet to be finalized.

5.3 Experiential Documentation Skills Help Improve Bottom-Up Initiative for Decision-Making

In follow-up interviews conducted in 2016, the utilization of documents as administrative tools to communicate and deliver information, including and negotiating with others, was particularly noteworthy. An irrigation schedule was implemented in the 2016 dry season based on proposals made by the P3A federation, starting downstream with gate opening/closing at focal locations. The federation leaders utilized examples highlighted in the handbook, especially the Appendix. Examples of useful documents created during the action research are given in Figs. 8.9 and 8.10. Most were made ad hoc and coincidentally.

Figure 8.9 shows the list of names and phone numbers of federation leaders and gatekeepers in Kampili region. It was created as a result of a meeting held in Kampili

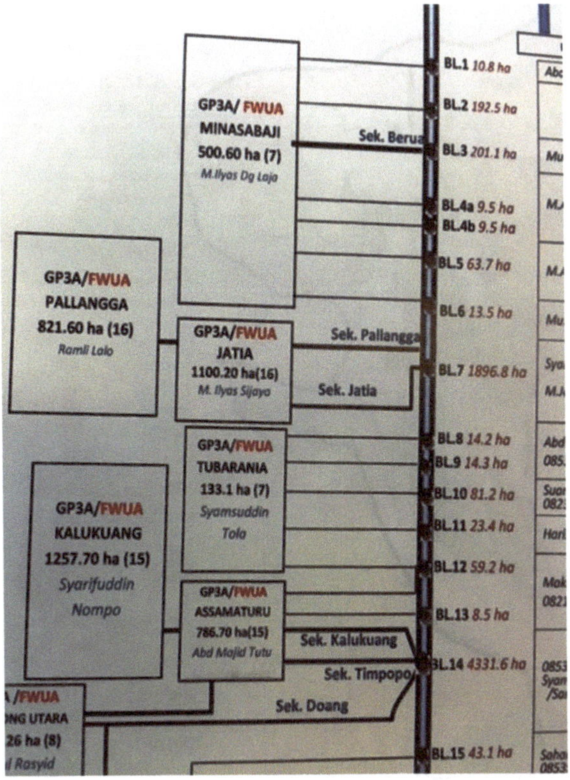

Fig. 8.9 An excerpt from the handbook (name and phone numbers) (Photo taken by the author)

in 2015 during which, by chance, a P3A federation leader asked for the phone numbers of the other attendants. Accordingly, everyone wrote their phone numbers on the board and asked local researchers to copy them into a document. Participants subsequently discussed who should have access to the phone numbers to facilitate daily work. Thus, much of the documentation filed in the handbook appendix was not planned in advance but created as needed and as a result of interaction. It is therefore the hope that in order to determine "what extent action, based on research, really resolves problems" (Edelenbos et al. 2010) can be observed through the ongoing and developing utilization of these documents.

Reaction from the governmental side was also observed. A provincial staff member, who is assigned to supervise the Bissua, neighboring area of Kampili, joined the action research in 2015 based on a request by the P3A federations in the area. The staff member happily reported that her governing area was shortlisted for the provincial prize for irrigation management in 2015. She realized that one of the common points among the top five candidates was the proactive involvement of the prefectural and village authorities in it. She was highly motivated and is willing to apply for the prize again by demonstrating some utilization of this handbook. She also seems to have gotten a more concrete image of her own role by joining in the action research, and it encouraged her ownership of her responsibilities in collaboration with farmers.

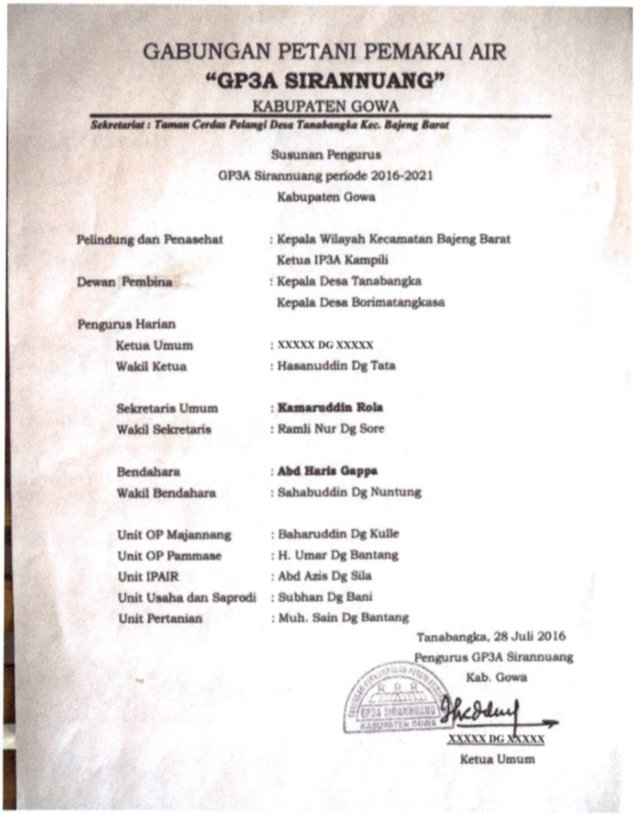
GABUNGAN PETANI PEMAKAI AIR
"GP3A SIRANNUANG"
KABUPATEN GOWA
Sekretariat : Taman Cerdas Pelangi Desa Tanabangka Kec. Bajeng Barat

Susunan Pengurus
GP3A Sirannuang periode 2016-2021
Kabupaten Gowa

Pelindung dan Penasehat	: Kepala Wilayah Kecamatan Bajeng Barat
	Ketua IP3A Kampili
Dewan Pembina	: Kepala Desa Tanabangka
	Kepala Desa Borimatangkasa
Pengurus Harian	
Ketua Umum	: XXXXX DG XXXXX
Wakil Ketua	: Hasanuddin Dg Tata
Sekretaris Umum	: **Kamaruddin Rola**
Wakil Sekretaris	: Ramli Nur Dg Sore
Bendahara	: **Abd Haris Gappa**
Wakil Bendahara	: Sahabuddin Dg Nuntung
Unit OP Majannang	: Baharuddin Dg Kulle
Unit OP Pammase	: H. Umar Dg Bantang
Unit IPAIR	: Abd Azis Dg Sila
Unit Usaha dan Saprodi	: Subhan Dg Bani
Unit Pertanian	: Muh. Sain Dg Bantang

Tanabangka, 28 Juli 2016
Pengurus GP3A Sirannuang
Kab. Gowa

XXXXX DG XXXXX
Ketua Umum

Fig. 8.10 A document created by the P3A federation for renewal of committee members (Photo taken by the author)

Figure 8.10 shows a document aimed at legitimization of a new P3A federation leader, created after termination of the research project. The federation lost one leader as a result of his sudden death in 2015, causing concern among the research team since a long wait to appoint a new leader after the loss of a committee member usually results in weakening of rural organizations. From the perspective of research, it was also a useful index for assessment of organizational capacity. However, the secretary of the federation (age 39 years old) noted that "This handbook consists of our own experiences. So we want to refer to the contents repeatedly." Accordingly, the federation took steps to consult with the weir keeper regarding how to proceed with formal meeting preparation, and, as a result, four village mayors in the area were invited to a meeting along with county servants, regency officials on agriculture and water resource management, and 90% of the P3As (of a total of 14 P3As) in the area. They utilized the opportunity to discuss additional matters, and donations were made by the village mayors to facilitate the large-scale meeting.

Action research in this area therefore encouraged the sharing of simple information such as lists of names and phone numbers, both of public workers in the field and representatives of farmers' organizations, providing opportunities for them to meet face to face and record and document discussions. The co-researcher of the Sulawesi team pointed out that the simple task of carrying out data collection was a new perspective in action research. By recording the process, and giving regular feedback to the farmers, the importance of basic, localized rather than generalized information in a rural setting was realized. Nevertheless, the task was challenging from the point of view of working as both a researcher and an NGO facilitator, and it was therefore concluded that care should be taken when collecting scientific data in such rural settings, because the research also becomes a commitment into people's lives.

6 Conclusions: Contextualizing Forms of Interaction

This study shows how multiple stakeholders facing discontinuity over each role and responsibilities can reconnect beyond cognitive boundaries by internalizing organizational values in the existing local context. The effectiveness of the action research was observed through the promotion of interaction between the various participants including MJs and government-led P3As and conventional P3A federations and often indifferent village officials. It also helped clarify the often confusing roles of the regency office and farmers' organizations. Although partially factual, some opinions regarding stakeholders were based on prejudice, creating cognitive boundaries to communication. From this point of view, concrete information such as making available the names and phone numbers of each participant, and providing opportunities to build a common area for daily communication, such as local meetings, becomes a simple yet important tool.

In addition, some participants, both farmers and government workers, are linked or overlap in their self-identities. One elite farmer working as a public servant was able to connect farmers and the village mayor, possibly even the ability to negotiate with county officials. In contrast, many gatekeepers and technicians, who are also public servants, have a self-identity as villagers. As mentioned earlier, the actualization of potential leaders throughout the action research period is therefore important. Most have a social/public role and responsibility within the village community (school worker or temporary village or county official in addition to working as a leader in a P3A federation). Such participants are key to building bridges between those facing discontinuities.

The government and citizens are generally assumed to be at opposite ends of a spectrum, at least at the beginning of interactions (Torfing et al. 2012). Residents in the field often represent a local historical perspective and, as a result, tend to feel distant from the government. However, current realities suggest that key participants can work to bridge the relationship through the interactive processes (Edelenbos 2016; Otsuka 2016). Under the setting of action research, the involvement of local researchers as mediators is also important in strengthening these key relationships. As shown in the results of the interviews, local researchers consciously took on the role of bridging stakeholders.

Discontinuity between P3As and MJs, as observed at initial interviews in 2011–2012, seemed to have occurred due to the lack of opportunity to jointly exchange values and institutionalized rules (such as new laws governing water resource management) between organizations (such as P3As and federations). As a local system, MJs were created during the lordship era, becoming embedded in a shared view of life, with local stories of water management having been handed down from parents to children. In this case, the action research defined a common space for collaboration between both farmers and government-led modern institutions such as the P3A and the P3A federation, helping link modern and historical settings.

Acknowledgment Fieldwork for this study was jointly conducted with the support of Dr. Dorotea Agnes Rampisela and her NGO, "Pelangi." The author is deeply grateful to the Research Institute for Humanity and Nature and the co-researchers of the joint research project (2011–2015) for their permission to use the data in this paper. All discussion within this paper is, however, entirely the responsibility of the author.

References

Assegaf, Abd Kadir. 2008. *Kitab Lontara Syekh Yusuf*. Makassar: Pustaka Refleksi.

Edelenbos, Jurian. 2016. Introduction: Three Reflecting Perspectives on Interactive Governance. In *Critical Reflections on Interactive Governance, Self-organization and Participation in Public Governance*, ed. J. Edelenbos and Ingmar Van Meerkerk. Cheltenham: Edward Elgar.

Edelenbos, J., N. Schie, and L. Gerrits. 2010. Organizing Interfaces Between Government Institutions and Interactive Governance. *Policy Sciences* 43: 73–94.

Kulle, Dg., Syarifuddin, and Zainuddin Tika. 2008a. *Aksara Lontara Makassar 1*. Makassar: Pustaka Refleksi.

Kulle, Dg., Syarifuddin, and Zainuddin Tika. 2008b. *Aksara Lontara Makassar 2*. Makassar: Pustaka Refleksi.

Long, Norman. 2001. *Development Sociology, Actor Perspectives*. London/New York: Routledge.

Long, Norman, and A. Long, eds. 1992. *Battlefields of Knowledge. The Interlocking of Theory and Practice in Social Research and Development*. London: Routledge.

Otsuka, Kenji, ed. 2016. *Interactive Approach to Water Governance: Case Studies in Asia (Interim Report)*. Tokyo: Institute of Developing Economies, Japan External Trade Organization.

Torfing, Jacob, B. Guy Peters, Jon Pierre, and Eva Sørensen. 2012. *Interactive Goverance: Advancing the Paradigm*. Oxford: Oxford University Press.

Yamori, Katsuya. 2010. *Action Research, Practical Human Science*. Tokyo: Shinyosha.

Kazuko Oguni is Professor of Development Anthropology in the Faculty of International Welfare Development, Nihon Fukushi University. Her research interests are in the fields of rural and agricultural development processes, cultural values among people, and the changes that take place therein. She has conducted research in South Sulawesi of Indonesia, in northern Cambodia, and in Japan. She was part of a multidisciplinary research project entitled "Designing Local Frameworks for Integrated Water Resources Management (2011–2015)."

Chapter 9
Interactive Governance at Anasagar Lake Management in India: Analyzing Using Institutional Analysis Development Framework

Mansee Bhargava

Abstract The governance of Anasagar toward its sustainability is driven by the physical, institutional, and community characteristics and their interrelations. They are crucial to document and understand the interactive water governance process in the urban lakes of India. Anasagar is located in the heart of pilgrim city of Ajmer city in Rajasthan India. The lake has been historically a rainwater harvesting reservoir; however, today not only rainwater but also wastewater is running into the lake. In the recent lake restoration process, the decided full tank water level of the lake has led to submergence of several developments at the lake shore. This has led to dissatisfaction among the local people since several of the submerged lands belong to them. Anasagar is an atypical lake in India whose lake area is divided into several land parcels and ownerships including both private and government. The water pollution continues in the lake since the inlet drainages carrying rainwater and wastewater from the surrounding hills and the urban development are awaiting treatments before flowing into the lake. Facing the dilemma of development and conservation of the lake by the physical, institutional, and community factors, the chapter discusses the complex problems and processes that influence the lake governance toward sustainable development.

Keywords Interactive water governance · Anasagar · Urban Lake · IAD Framework

1 Introduction

Local lessons are useful for theoretical foundation for the growing interactive governance (Torfing et al. 2012) in India and in Asia after it has picked up in the Europe especially in the field of water governance (Edelenbos and Van Meerkerk 2017). In

M. Bhargava (✉)
Environmental Design Consultants, Ahmedabad, India
e-mail: mansee@edc.org.in

K. Otsuka (ed.), *Interactive Approaches to Water Governance in Asia*,
https://doi.org/10.1007/978-981-13-2399-7_9

this paper, the state of interactive governance in India is evaluated through a case study of Anasagar[1] in Ajmer. The focus of interactive governance is on collectivity of governments, private, and societal actors with activities like participation, civic initiatives (Edelenbos 2005), self-organization, information sharing, and communication (Ostrom 2009). That way the notion of interactive governance is a part of more holistic approaches used in managing water resources in Asia such as integrated water resource management (GWP-TAC 2000) and Integrated Lake Basin Management (ILBM) (ILEC website) where participation too is at the core. The ILBM is applied in the decade old National Lake Conservation Plan (NLCP) in India (MoEF 2005) that is a first national level planned restoration initiative for urban lakes, such as the Anasagar. ILBM endorses participation of the stakeholders as one of the key characteristics among the others, namely, policies, institutions, technology, information, and finance. Participation or no participation alias interactive governance is a derivative of a number of physical, institutional, and community characteristics that jointly affect and are affected. These characteristics are well encompassed in the institutional analysis development (IAD) framework (Kiser and Ostrom 1982) as an input source for the desired and evaluated outcomes like the interactive governance. The framework is used to analyze the characteristics that drive the interactive governance in Anasagar. The contextual and institutional characteristics, interactions, outcomes, and evaluative criteria along with the direct and feedback links form the structure of the framework.

In the last decade in India, on the one hand, urban lakes have gained prominence in the environment-development policies, and on the other hand, their sustainability is confronted by the increasing societal demands of services from them and weak governance approach to sustain them as well as meet the societal demands (Bal 2015a, b). Though urban lakes are still in abundance in India, there has been a systematic deterioration of urban lakes in terms of numbers as well as size. The systematic deterioration is a governance outcome resulting from a number of physical, institutional, and community developments over time. Anasagar is a classic case of how the physical, institutional, and social environments facilitate participation under certain circumstances and how otherwise the participation is weak. Anasagar is located in the heart of Ajmer city in Rajasthan, India. Ajmer is a pilgrim city with the famous *Dargah*[2] *Sharif* located in the vicinity of the Anasagar. Nested amidst a picturesque hilly landscape, Anasagar holds a religious and cultural importance for the local people as well as the pilgrims across India who visit the *Dargah Sharif* throughout the year and is peaked during festivals especially Urs. The social and cultural characteristics of the community (local and the pilgrim) play crucial role in the governance of Anasagar. Though historically a rainwater harvesting structure (Agarwal et al., 2001), today Anasagar is a combined rainwater and wastewater reservoir. The lake was heavily polluted when the NLCP project for Anasagar was

[1]Anasagar in Hindi is made of two words: Ana is a name of the ruler Anaji Chauhan during whose rule the lake was constructed by the people of Ajmer and *Sagar* means water body. From hereafter, the term either Anasagar or lake is used.

[2]*Dargah* is a tomb of a Muslim intellectual who is considered "holy" by the Muslim community.

initiated in 2007 by the local planning organization of Ajmer, namely, the Ajmer Development Authority (ADA) with the support of the Government of Rajasthan (state). The ADA delegated a consultancy to craft a road map for an integrated restoration plan of Anasagar based on the ILBM guidelines. A draft project report (hereafter referred as Anasagar restoration report) on an integrated restoration plan of Anasagar is prepared (IL&FS 2007). The funding, implementation, management, and maintenance activities at the Anasagar began in 2008 as per the Anasagar restoration report (ADA 2007). A key component of the report is the participation from the different stakeholders in the different lake activities, i.e., a sort of interactive governance.

This chapter looks at the physical, institutional, and social environments of Anasagar and Ajmer that are driving the governance of Anasagar with the implementation of the NLCP. The enquiries are: Which and how the physical, institutional, and community characteristics influence governance of Anasagar toward its sustainability? And what is the observed pattern of the governance? The main body of the chapter involves categorically describing and explaining the physical, institutional, and community characteristics and their linkages, those that are influencing the interactive governance, and the key aspects defining the patterns of interactive governance. The study is a part of an ongoing effort on developing a database of urban lakes based on a social-ecological system approach focusing on the generic characteristics of success/failure of lake governance and sustainability (Bal 2015a, b) toward building ontology of urban lake system studies.

2 Conceptual Research Framework

Participation of the stakeholders is a common link between the notions of interactive governance, the ILBM, and the IAD. A conceptual research framework constituting the three is shown in the Fig. 9.1 and is discussed here with reference to Anasagar.

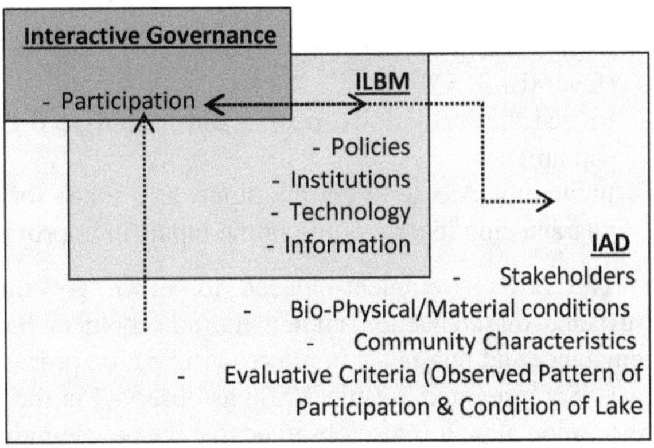

Fig. 9.1 Research framework (Source: the author)

The interactive governance (Kooiman 2003), in its traditional avatar, exists in India since antiquity referring it to community-based management (Agarwal and Narain 2001) and collective action (Ostrom 1990) toward managing natural/common pool resources like lakes, forests, rivers, etc. The literature primarily then was about managing small-sized resources by small homogenous groups mostly in non-urban settings. With large-scale resources like the Anasagar, it was a decision of a ruler that was agreed upon and carried out by the local people. Experts like Anil Gupta and Nimish Patel (in conversation in interviews, 2014) working on the traditional wisdom on water management in India argue that construction and management of the traditional water harvesting structure like Anasagar is through collective governance where the decision of the ruler was a result of the interactive processes between the then local people and experts and management was mostly the responsibility of the local people who were also the users. The multiple ownership of the land encompassing the Anasagar further supports the argument as discussed later.

The rise of interactive governance (Peters and Pierre 2012) in the recent time focuses on the participation, interactions, and initiatives of the multiple (government, private, and people) actors in dealing with the complex social-ecological systems (Berkes et al. 2003). There is, in the hindsight, a question about real participation of the local people in urban contexts and cases like the Anasagar management such as what is the mode of participation (Arnstein 1969); who induced the participation; what are the conflict resolution mechanisms (Ostrom 1990); importantly, why do people participate or not participate; how is participation manifested; etc. Participation of the local people in the interactive governance is manifested in different ways across different disciplines, sectors, and regions, for example, civic engagement (Putnam 2000), stakeholder participation (Arnstein 1969), self-organization (Ostrom 1990), civic initiatives (Edelenbos and Van Meerkerk 2017), partnerships (Pierre 1998), etc. It is noticed that in urban settings and particularly in large-sized resources like Anasagar, the interactive governance is strongly organized by the governments, meaning government often decides when, who, and how people get involved. There, at certain moment in decision-making process like the consultation in preparation of the Anasagar restoration report, the ADA invited peoples' participation for inputs in highly structured settings of rules (dos and don'ts) and with their stoned ideas that there were these situations:

- Invited the preferred network of people who will anyway approve the ideas (favoritism)
- Invited the well-known people and incentivized them to favor the ideas (elite capture)
- Invited the people to get the inputs as a token for people support however not guaranteeing incorporation of the inputs in improving the ideas (tokenism)

The above government-induced (top-down) governance situations create lack of trust and dissatisfaction among the stakeholders (Edelenbos 2005) resulting in reluctance and resistance in future participation processes of representative democracy (Sørensen and Torfing 2007) as observed in the implementation of Anasagar restoration plan. It is also observed that the people-induced (bottom-up) governance situations are time-consuming involving higher transaction costs and with weaker control over decision-making and implementation; however, groups of people have taken on initiatives that address, collaborate, and resist the current lake governance through a number of activities as discussed later.

To address the above situations, it is necessary that the discourse of interactive governance in complex social-ecological systems is more than the collectivity of people and organization. Focused on lake management, the ILBM encompasses participation as one of the characteristics among the others, namely, policies, institutions, technology, information, and finance. The ILBM (from the International Lake Environment Committee, ILEC Foundation of Japan) is a mandatory guideline for the conservation and management (c&m) of the urban lakes under the NLCP[3] (MoEF 2005). According to the ILEC (2007), the ILBM or the lake basin governance is possible by integrating the abovementioned characteristics; ILEC provides normative ideas about these characteristics. *Participation* of the multiple actors is about defined position and role in the management decision-making process. The local government and the local community are de facto vital actors. *Policy*, as rule of the game, is about the tools that facilitate concerted social actions for sustainable lake basin management. *Institution* is about an organizational setup that works toward the sustainable benefits of the resource users. *Technology* is about the physical intervention into the lake such as provision of wastewater treatment system, afforestation, desiltation, shoreline and wetland restoration, etc. *Information* is about generation and sharing of knowledge to mobilize human and financial resources and to minimize the difference in perceptions about lake management between science and people. *Finance* as a resource is about deriving innovative approaches to generate locally usable funds from all basin stakeholders benefiting from the resource values of direct and indirect uses of a lake.

The ILBM is a process and not a part of project that ends at a defined period. The process involves regularly documenting the present reality to envision the future possibility and in between involves monitoring, database development, envisioning the future governance, and reviewing and assessing the improvements (Lintner 2007) (refer to Fig. 9.2). A long-term and short-term planning to carry out the above is crucial. A challenge with the ILBM approach in Anasagar is that it was seen as a project duration thing and a panacea for problem-solving. It is found that after the completion of the NLCP, Anasagar is still challenged with a number of aspects pertaining to the long-term and short-term planning for its restoration.

To analyze the performance of the interactive governance with the ILBM characteristics, it is necessary to have guiding causal relations that help in explaining the inputs, the process, and the criteria to evaluate both. The IAD (Kiser and Ostrom 1982) is a useful framework to understand the causal relations between the ILBM characteristics with respect to participation. Its structure originates from the general systems approach where inputs are processed by the stakeholders (through

[3] The concept of ILBM started in India as an international agreement of knowledge and technology exchange through the JICA projects on integrated water resource management piloting with the Bhoj Wetland Project and followed by a number of other wetland restorations.

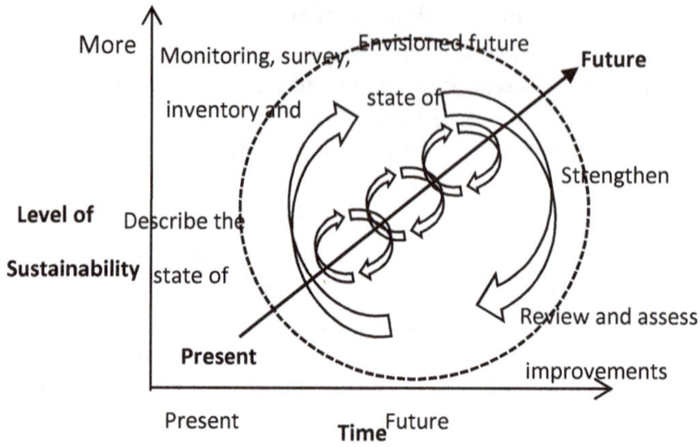

Fig. 9.2 Conceptual illustration of ILBM cyclic process. (Source: Lintner 2007)

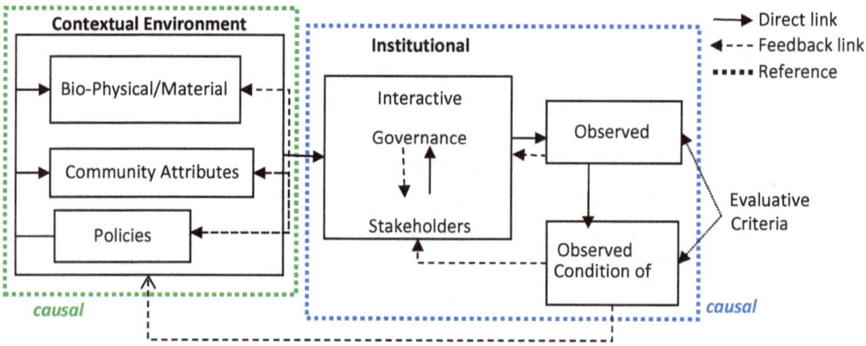

Fig. 9.3 Contextualizing the IAD framework (Source: Adapted from Bal 2015a, b from Ostrom 1990)

interactive processes) into outputs that have outcomes (both governance and physical) that are evaluated with feedback effects (McGinnis 2011). In the framework, the input criteria are the ILBM characteristics besides the physical characteristics of the resource, the process is about the stakeholders in the interactive governance activities, and the evaluative criteria are for both the governance and the lake performance/conditions. Figure 9.3 presents the basic structure of the framework. The conceptual representation of the ILBM process over time in Fig. 9.4 fits the framework in a linear form.

It is a microeconomic analytical method (McGinnis 2011) for analyzing institutional designs like the interactive governance toward managing natural/common pool resources (Ostrom 1990) like the urban lakes. The contextual environment is an independent variable that includes the biophysical characteris-

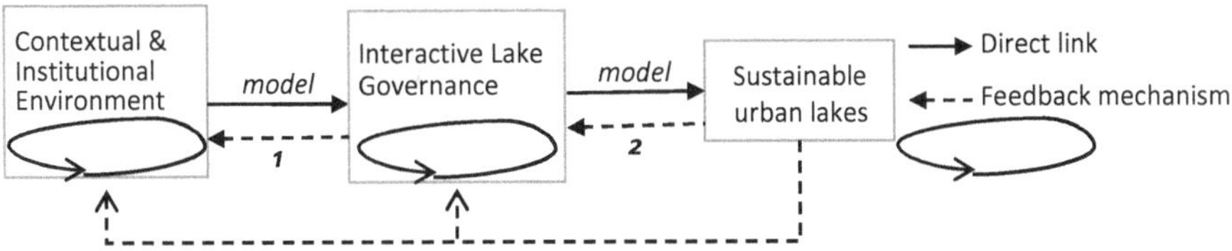

Fig. 9.4 A model representing the analytical causal relations (Source: Adapted from Bal 2015a, b)

tics of the lake and the community attributes and the policies; and the institutional environment is the dependent variable that includes the stakeholders and the governing activities. The latter is the subject of analysis, and the former sets the background of the analysis. The biophysical condition refers to the size, number, location, biodiversity, and human-constructed facilities of the resource. The material condition refers to the nature of use of the resource. The nature of resource is primarily determined by the physical characteristic of the resource for/after its use (Ostrom 2005). The stakeholders participate directly and indirectly in the governance activities, such as policy-making and planning of the lake activities, implementing and maintaining the lake, and even using the lake functions. The local people using the functions of the lake or are benefitting from its existence are classified as the community. The polycentric lake governance (Narayanan and Venot 2009) of participation (Edelenbos 2005) and collectivity (Ostrom 1990) of the governing organizations and the community in the lake restoration activities is what is referred as interactive governance. The outcome refers to the perceived interactive governance and observed condition of the lake sustainability. The evaluative criteria are the pattern (interaction) of governance and the physical improvement (outcome) of Anasagar.

The general system model used for analysis is manifested in a causal model (Bal 2015a, b) shown in Fig. 9.4. The enquiry is framed as which and how the physical, institutional, and community characteristics and their interrelations drive the interactive governance of Anasagar toward its sustainability. The enquiry is also on the observed pattern of interactive governance. The focus of analysis is on explaining the model 2, and in the process, the model 1 is described. Implicit in the causal model is that interactive governance drives lake sustainability. This normative assumption is also found in the Anasagar restoration report (IL&FS 2007). The outcome and evaluative criteria of participation and improved lake condition are also in the alignment with the objectives of the lake governing organizations of Ajmer (ADA 2007).

The analysis below is descriptive and explanatory. It is based on the acquaintance of the subject and the place; observations and information extracted from papers, official reports, and media; and responses from the open-ended interviews of the local government officials and the local people. The analysis and the findings become a contribution to the ongoing database preparation (by the author) on urban lake governance and sustainability in India.

3 Anasagar in Ajmer

Anasagar is located in the city of Ajmer. The geographical location of Ajmer with Delhi and Jaipur is significant to its 1400 years of historical events that shape the city's urban morphology. It is here: the shrine of the Sufi saint, Khwaja Moinuddin Chisti, famously called "*Dargah Sharif*" laid the foundation of Islam in India in 1192 AD; and the East India Company laid the foundation of the British Raj in India in 1616 AD, through the charter of free trading granted by the Mughal Emperor Jahangir (AMC and PMB 2006). Anasagar is a historical man-made lake built by constructing an embankment across the river Luni between the two hills, namely, Bajrang Garh and Khobra Behrun, during 1135–1150 AD under the rule of Anaji Chauhan, and thus got its name as Anasagar.

Anasagar comes within the Ajmer administration and is managed through multiple stakeholders holding a complex administrative structure for the c&m of the lake. Ajmer city is the district capital of Ajmer district that is headed by the district collector/district magistrate (DM) for all matters including Anasagar. The two other cities that form a tri-city with joint administration in the district are Pushkar and Kishangarh. The planning and development of the tri-city is with the Ajmer Development Authority (ADA). The Ajmer-Pushkar joint Municipal Corporation (AMC) is involved in the maintenance of the city infrastructures including the Anasagar. People residing in the three cities have strong association and perception about Anasagar besides the other lakes of high importance, namely, Bada Pushkar in Pushkar and Gundalav Talav in Kishangarh.

Ajmer is an important religious city of the country. The *Dargah Sharif* is the main attraction for pilgrims from all religions. Pushkar hosts the only Brahma Temple in the country and is another attraction for the pilgrims. Millions of pilgrims and tourists visit them every year. The peak visitors' inflow at the *Dargah Sharif* is during the Urs festival, the Eids, and the Ramazans; and at the Brahma Temple, it is during the Pushkar *Mela*[4]. In addition, Kishangarh is the marble trading capital of India where there is a regular flow of business visitors. People visiting the three cities visit Anasagar for various ritualistic and recreation purposes.

The city of Ajmer is undergoing planning as per the Ajmer Development Plan (2013) prepared for the Ajmer Metropolitan Area (AMA). The total AMA land of 75,751.56 ha is predominantly covered by agriculture, forest, and mountains. The main economy of the area is agriculture and tourism, and Anasagar has a vital role in tourism and in agriculture in the downstream. The population of AMA is above 550,000 (551,360 as per Census, 2011). Around 30% of the total population resides in the catchment area of Anasagar. Besides, Ajmer hosts a large floating population of pilgrims and tourists that even reach nearly quarter of the total population during

[4]*Mela* is a Hindi word that means fair. The Pushkar Mela is celebrated for 5 days in the month of October to November during the full-moon days. It is celebrated as the birth of Pushkar Lake by Hindu god Brahma; thus numerous people take a dip in its sacred waters. It is famous also as a livestock fair, especially the camels. As a popular event, millions of tourists across the world visit Pushkar at this time to experience the vivid culture of Rajasthan (http://www.pushkarmela.org/).

the month-long Urs festival. While the local people feel strongly associated with Anasagar and take pride in it besides being concerned about it, the visitors consider Anasagar as sacred, and several of them perform holy dip in it as a belief to overcome diseases and problems of life.

4 Contextual Environment

Anasagar is typical traditional rainwater harvesting structure (Agarwal et al. 2001) found in India. The biophysical and material conditions of the lake are broadly characterized by the social-ecological conditions of the lake, the lake shore, and the lake catchment area, together covering a geographical area of approximately 56 km^2 (IL&FS 2007).

The city is situated in the cradle of the Aravalli mountain range and surrounded by Nag hills, Madar hills, and Taragarh hills. Nested within the mountains, Anasagar is located on the North East Valley (Fig. 9.5). There are other water bodies in the valley, namely, Foysagar, Paal Bichla, and Khanpura tank; and a number of natural drains, together with Anasagar, constitute to around 375.03 ha (0.49%) of the total AMA area. Several livelihood activities are attached to the forests and the water bodies.

The delineated area of Anasagar is ambiguous since some land parcels at the lake shore are under legal conflict. Depending on the water level in the lake, the full tank area varies between 97 ha and 182 ha; the circumference varies from 4.8 km to 7.3 km; and the depth ranges from 1.9m to 4.4 m (IL&FS 2007). The full tank capacity of the lake in the early 1900s had a circumference of 12.87 km (Sarda 1911).

Anasagar shore is surrounded by diverse activities building an image of the lake as well as the city. The lake is surrounded on three sides by a road, namely, *Gaurav Path.*[5] The southern part of the lake is surrounded by the *Pushkar* Road (a part of the National Highway 9) that connects the tri-city. The roads abut a number of historical recreational, commercial, institutional, and residential developments. The southern side of the lake shore has *ghat*[6] and gardens. On the south-east end, the constructed overflow dam is a 335-meter-long stone embankment. It has four outflow sluice gates. The historical garden *Daulat Bagh*[7] (garden of splendor) abuts the dam. *Daulat Bagh* was built under the Mughal Emperor Jahangir in the early seventeenth century. Jahangir also built a palace (*Mahalat-i-Jahangiri*) here adjoining the *Kaiser Bagh*, the remnants of which do not exist anymore. Later, in the mid-seventeenth century, the marble pavilions called *Baradaries* were built under Shah Jahan. For

---

[5] *Path* is a Hindi word referring to road or a path.

[6] *Ghat* is a Hindi word referring to stepped embankment that is used by people for various physical interaction activities with the water like bathing, praying, washing, etc.

[7] *Bagh* is an Urdu word of garden.

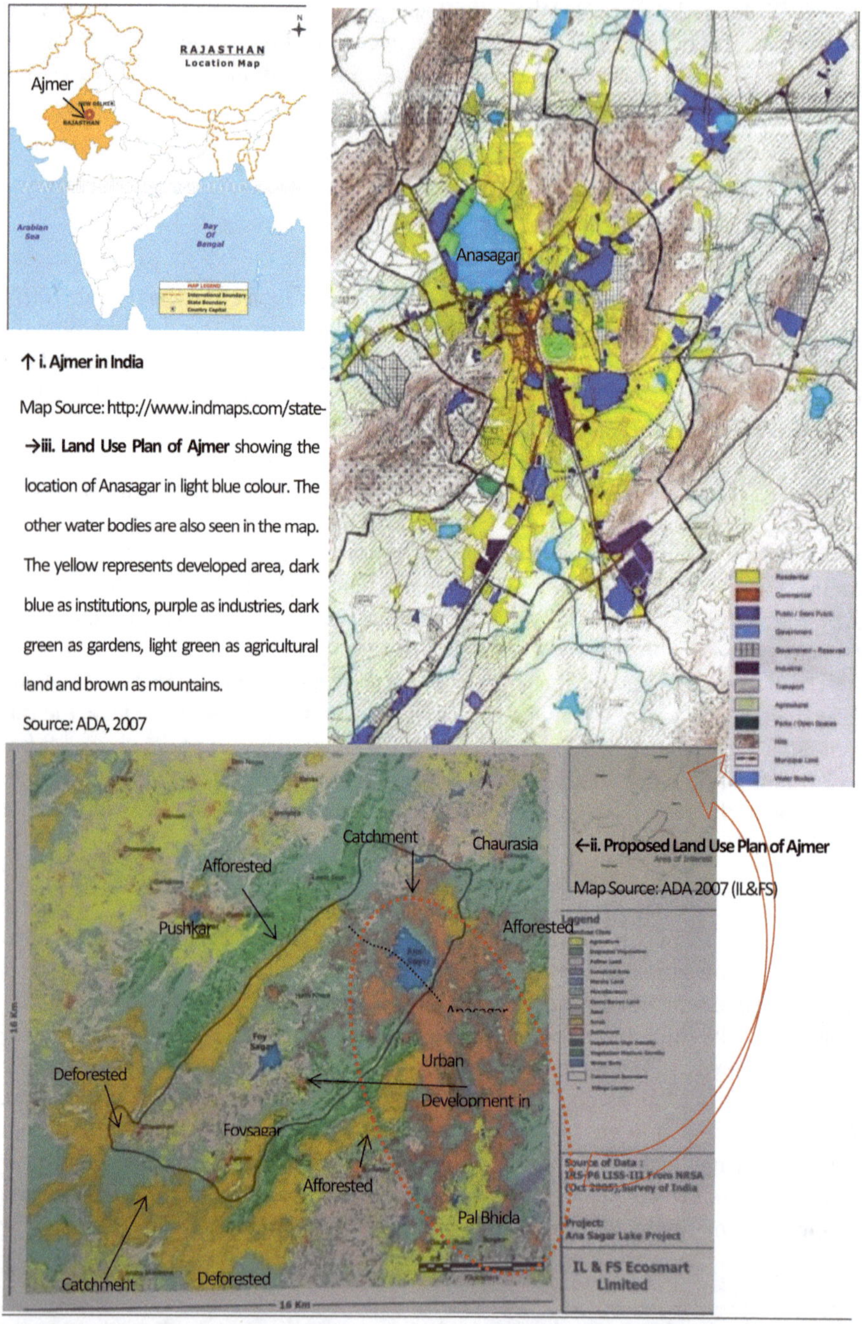

Fig. 9.5 Urban morphology of Ajmer

the Mughal rule until part of British rule, these pavilions were housed with different administrative purposes (Sarda 1911).

On the east shore is a designed lake promenade, namely, *Chaupati*,[8] constructed during the NLCP that was sanctioned in 2008. It is the heart of social activities in the city. Local people visit *Chaupati* on regular basis for various activities such as fitness, socializing, resting, and meditating besides enjoying the picturesque view of the lake especially during the sunrise and sunset. The Ramprasad *Ghat* in the south side on the Pushkar Road provides physical access to the lake. People interact with the lake in different ritualistic and recreational ways here such as bathing, worshipping, washing, swimming, fishing, fish feeding, etc. Pilgrims visiting the *Dargah Sharif* consider a dip in the Anasagar to be holy. Several pilgrims do so apart from taking a bottle of water from Anasagar as holy water. Here, the Pushkar Road, the *Ghat*, and the lake full tank level are almost at the same level. Between the *Chaupati* and Ramprasad *Ghat*, a small and highest hill point houses the Circuit House (a political house of the district) and a Hanuman Temple offering a panoramic bird eye view of the lake and the lake environs (Fig. 9.6). At the foothill of the Circuit House near *Chaupati* is located a Dhobi (cloth washing) *Ghat* that local people consider as one of the sources of lake pollution.

There is an island near to the south-east side of the lake closer to the *Daulat Bagh*. It is developed under the NLCP and hosts a recreational park with food joint and children play facilities. The island is also a home to several aquatic birds. A boating jetty located at the *Daulat Bagh* is used to bring people to the island besides recreational boating in the lake. A number of water fountains with light effects are installed in the lake between the island and *Chaupati* as part of the NLCP that are a source of attraction in the night. The *Daulat Bagh*, *Chaupati*, Ramprasad *Ghat*, and the Hanuman Temple together form the prominent lakefront public places and heart of the city's social-cultural activities.

At the south-west edge is a fishing bank. The state fisheries department has contracted the fishing activity to a private contractor for fishing once a day. A private landowner has leased out the place to the contractor to use it for boat parking, fish landing, washing, and packaging for dispatch to the vendors. According to the contractor and fishery officials, there is room for more fish catch from Anasagar owing to high fish population; however fishing is restricted as per the Anasagar Restoration Board (hereafter referred as the Board). There is a growing concern about the increase of African carb in the lake that is negatively affecting the lake ecosystem.

On the south-west side of the lake, a secondary treatment plant and an artificial wetland are constructed under the NLCP. They were found submerged in the lake water level in the number of field visits made in different times of the year. These

[8] *Chaupati* is a Hindi name referring to a waterfront with various recreational activities especially food joints and games. It was first coined for the Mumbai seafront, and since then waterfronts in several cities across the country have used the name.

Fig. 9.6 Anasagar: catchment, shore, and the basin

Fig. 9.7 Submerged buildings at Anasagar (Source: Ajmer Patrika, 06 Oct 2014)

wastewater treatment structures are yet to start functioning (as on March 2016). Local experts even argue that the capacity of the treatment structures is lesser than the total wastewater inflow, and, therefore, even if they start functioning, they may be inefficient. On this side of the lake, there are also a number of submerged buildings owned by the local people (Fig. 9.7). The cause of submergence is attributed to the increase in the full tank level of the lake under the NLCP. A newspaper article of October 2014 (Appendix 1) summarizes the submergence issues at Anasagar. Similarly, the planned developments (such as Gulmohar Park, Vaishali Nagar) at the lakefronts in the north-east side inside the Gaurav Path face threat to flooding. The developments here are built on the land reclaimed from the lake submergence area in the previous city planning in the 1970s. The plinth of the buildings here are below the full tank level of the lake leading to backwater inflow from the lake into the buildings. A wall was constructed (in the 1980s) at the lake edge to protect the developments from backwater inflow that resulted later in a constructed wetland situation here receiving wastewater from the developments. The backwater flow and water seepage are also a concern for the buildings here. The stagnant water in both these areas are considered key source of waterborne health problems for the local people, according to the medical practitioners of the city. Local people living in these areas are in a legal battle with the local governments regarding reducing the full tank level of the lake.

The north-west side of the lake facing the Gaurav Path houses the Regional Education Institute campus that to large extent protects the land around it to further subdivisions for buildings. Across the road facing the lake are the few land parcels belonging to private owners. The Gaurav Path and the full tank level of the lake here are almost at the same level. This is a stretch of the lake shore that has open access to the lake; however there are no public activities seen here due to the safety reasons as per the local people.

The gross catchment area of Anasagar is 56 km^2 of which about 30% is covered by urban development. Historically, all the slopes from the three hills, namely, Nag hills, Madar hills, and Taragarh hills, formed the natural catchment area of Anasagar. They set the picturesque background for Anasagar besides protecting the ecological balance of the area. The catchment area of Anasagar diminished first with the construction of Foysagar (near the hills of Ajaipal) during the British rule (in 1891 under engineer Foy) and further with the urbanization around the lake and in the foothills. Several parts of the hills are deforested in the urbanization process. The remaining parts of the hills are now protected forest areas under the State Forest Department (SFrD). A number of afforestation programs including small check dam constructions in the hill slopes are carried out under NLCP to improve the biodiversity and to control the soil erosion through surface runoff from the hills. However, surface runoff from the hills and a regular wastewater flow from the developed areas carry a large quantity of silt and nutrients into the Anasagar and pollute it. Besides, both the developed and the underdeveloped areas in the lake catchment (Agrawal n.d.) witness heavy extraction of groundwater through bore wells and tube wells for domestic as well as irrigation purposes leading to groundwater depletion. Anasagar is an important source of groundwater recharge.

Anasagar along with the other water bodies Foysagar, Paal Bichla, Khanpura tank, and a number of natural drains, in the valley area of the mountains, forms an interconnected watershed system (Fig. 9.8). There are a number inflow drainages called *nala*[9] that carry rainwater and wastewater into the lake, some of which are constructed channels. The two main inflow drainages are the overflow from the Foysagar and Hathi Kheda *Talaab*[10] located at the south and flowing into Anasagar from the south-west and from the Chaurasia Talaab located at the north and flowing into Anasagar from the north. The overflow *nala* of Anasagar flowing out from the sluice gates at the Daulat Bagh becomes an underground drainage that is an inflow to the Pal Bichla *Talaab*. Most *nala* are found blocked at several places by the development of buildings and roads and solid waste accumulations that lead to water logging situation in the surroundings during the monsoon.

Historically, Anasagar supplied drinking water to the old city of Ajmer, before the construction of Foysagar, through two underground masonry channels, one passing through the city and the other outside it to the Surajkund near the railway station. These structures do not function anymore. Water supply in Ajmer is now from the Foysagar and Bisalpur dam. Anasagar is now primarily a natural, social, and historical entity to the city besides its natural function as water reservoir, groundwater recharge, and source of water for irrigation in the downstream villages around Pal Bichla.

The above state of the biophysical condition of Anasagar lake system highlights that the dilemma of governing a lake versus lake system prevails. A challenge in the urban lake governance is to address simultaneously the geo-spatial scales. Delineating a lake or a lake system for governance has different considerations and involves different mechanisms (Bal 2015a, b). While governing "a" lake involves

[9] *Nala* is a Hindi word originally referring to a natural drainage course or narrow stream of water. The term is now used for open and dirty drains either natural or constructed.

[10] *Talaab* is a Hindi word referring to small lake or pond.

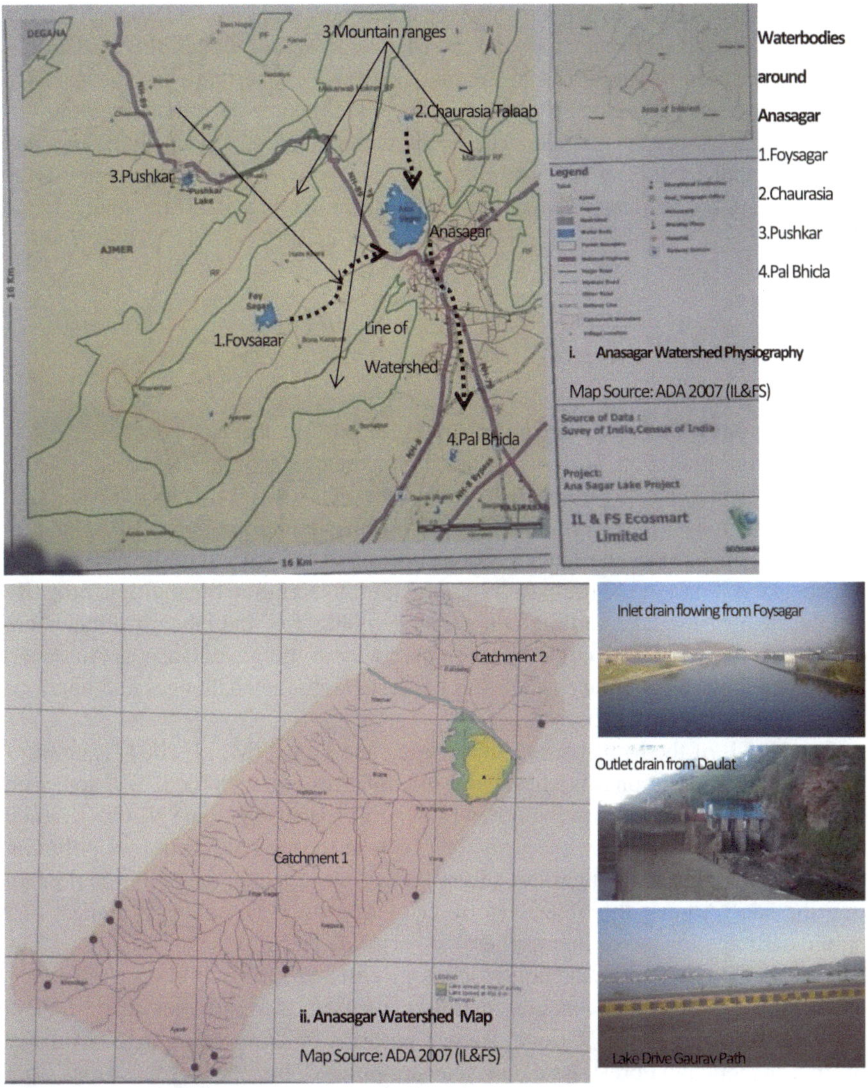

Fig. 9.8 Anasagar watershed

conserving of the lake submergence and protecting the lake shore, the catchment area is pressurized for urban development leading to re-delineation of the watershed, for example, about a third of the catchment area is covered by urban development. In addition, the land use planning of the catchment area (alias city) through the Ajmer Development Plan (2013 and before) poorly addresses the watershed planning and design of the lake and the shore. So, the current efforts at the lake and the shore end up becoming fragmentary to the challenges. For example, though the de-eutrophication through manual and mechanic processes under the NLCP has removed the water hyacinth from the lake, experts believe that even the repeated removal is temporary as long as there is inflow of large amount of untreated rainwater and wastewater into the lake. The development of public places like *Ghat* and *Chaupati* address the social needs but not the hydro-engineering needs of the embankment and drainage. Their poor maintenance in addition brings apathy to the lake shore.

5 Institutional Environment

A recent policy that influenced the Anasagar governance is the National Lake Conservation Plan (2007–2014). It is complemented by the Jawaharlal Nehru National Urban Renewal Mission that worked between 2004–2014 (JnNURM, 2005). While the former is based on the integrated lake basin management (ILBM) guidelines for c&m of lake, the latter is about integrated infrastructure development in which lake is a part of the drainage infrastructure. In addition, the multiple property right system of the lake and the bureaucratic volatility of the lake governance are crucial. These four are bundled in a complex way in the governance of Anasagar, and unbundling the complexity helps in understanding the challenges and opportunities of the interactive processes.

The NLCP of the Ministry of Environment and Forest (MoEF 2004) is conceptualized from the National Wetland Conservation Program with the ILBM approach to address the complexity in the urban (and peri-urban) lakes. The NLCP is aimed to restore the water quality and ecology of a lake through prevention of pollution from point source and interception, diversion, and treatment of the pollution flowing into the lake. Among the objectives (refer to Appendix 2), public awareness and public participation are of interest here.

The NLCP is a multi-stakeholder-level governance approach in terms of financial, technical, and human resource sharing in the lake c&m activities. The NLCP is implemented through the National River Conservation Directorate of the Ministry of Environment and Forest. As a policy, a committee of member organizations from state and local government was created as Anasagar Restoration Board (refer to Box 9.1). The Directorate is supposed to provide support to the Board in the form of funding (actual), technical, judiciary, human, and administrative. The Board has a mandate to submit annual work progress report to the Directorate as part of project monitoring and evaluation for approval of further funding and activities.

A shared funding arrangement was designed to facilitate participation of the stakeholder organizations in the implementation activities. The Directorate is supposed to support with 70% of the total proposed cost of lake restoration. The 30% share is divided between the state and the local government as 20% and 10%, respectively. The 10% share of the local government is to ensure periodic operation and maintenance of the lake functions and community participation, and the 20%

Box 9.1: Anasagar Restoration Board Organizations in the Anasagar Governance

The **National** River Conservation Directorate of the Ministry of Environment and Forest	
State – offices/ officials located in Jaipur	**Local** – offices/ officials located in Ajmer
1. Principal secretary (PS) to chief minister	8. Divisional commissioner
2. PS to Ministry of Urban Development and Housing, finance, PWD, PHED, and GWD	9. District collector
3. Secretary, local self-government development	10. Chief executive officer, municipal corporation
4. Secretary, planning department	11. Executive engineer, PIU, RUIDP represented by the ADA as nodal organization
5. Chief executive director, RUIFDCO	*As notified (no. 26775–90) on 19.11.2009
6. Project director, RUIDP	
7. Chief engineer (CE) PHED, PWD, irrigation	

Source: ADA (2007)

share of the state government is a cautionary fund to support other related activities and unforeseen events and disturbances (MoEF 2008).

The above institutional arrangement on paper is however a complex situation on the ground. First is the stakeholder. The state level organizations mentioned in the Board are merely a bureaucratic/governmental mandate. The role of the district collector is primarily monitoring the development and maintenance of the lake. The AMC and the ADA play the pivotal roles. The ADA was appointed the nodal organization through a state-level organization called the *Rajasthan Urban Infrastructure Development Project* for planning, coordination, and implementation; and the AMC is primarily involved in the implementation of the infrastructure through the JnNURM. The AMC is incharge of maintaining the rainwater-wastewater drainage systems, solid waste, and safety of the people linked to the lake. A number of other stakeholders not included in the Board but crucial and involved (in practice) in the lake governance processes are identified such as:

- The state irrigation department controls the water quantity (inflow and outflow). Irrigation and engineering works of a lake and its inlet and outlet drainage channels are constructed and built by the department. The department is the main authority to maintain the lake water quantity and the water level.
- The state fisheries department controls the fish population and fishing activities. Fishing and fishes are regulated through the terms and conditions set by the department. The department decides for the type of fish breeding in the lake. The contract for fishing is made with the contracting parties usually for 3 years and is subject to renewal based on the evaluation by the department.

- The state agriculture department takes charge of the crops and farming activities for the land parcels where the water level recedes and farming is allowed. Farming on the lake bed is regulated through the terms and conditions set by the department. The contract for farming is made with the contracting parties which in this case are the local landowners and is subject to renewal based on the evaluation by the department. Farming has stopped in the lake after the NLCP since there is year-round water in the lake.
- The State Pollution Control Board does the monitoring of water quality on regular basis. The office is located in Kishangarh. They have water sample points at different parts of the lake and the inlet/outlet drainage canals. They produce monthly and annual water health data and recommendations for addressing pollution that are shared with the Board.
- The political and bureaucratic influence in the Anasagar governance process is volatile. With every field visit, it was found that there was a new district collector and AMC commissioner. With every new party in power, the governing of the land-water dynamics of the lake changes. While the local bureaucracy is influenced by the state, the frequent changes make the decision-making and implementation slow and difficult. It is also a challenge in monitoring and evaluating the previous efforts since every time there is attempt to restart due to the personal priority of the leadership. The change in leadership also changes the other interactive processes including selection of the agencies for various activities.

There is delegation of responsibilities by the Board and the other organizations to large private organizations (which are at this moment not local) for the operation and maintenance of the lake-related activities that further delegate the responsibilities to smaller organizations or individuals locally for the implementation. The number of delegation of activities involves high transaction costs and misses the true value investment on the ground work. Besides, the overlapping roles cause unclear responsibilities and negatively affect the coordination and monitoring of the lake activities. There is a poor participation of the private businesses in the lake governance activities for various reasons including the property right system lake surroundings. In addition, the lake governance activities are not incentivized toward economic benefits as anticipated by the private businesses. However, there are several other models of participation of the private businesses applied across the country that includes adoption, BOT (build, operate, and transfer) of recreational activities, etc. (Bal 2015a, b). Similarly, the academic institutions are not directly engaged in the lake governance activities; however, there are a number of studies done on the social-ecological aspects of the lake. Pandey et al. (n.d.) compiled a list of publications on Anasagar in their report on "Evidence-Based Holistic Restoration of Lake Anasagar" submitted to the High Court of Rajasthan. In a workshop organized (as part of the RIHN Fellowship) on "Water and Workshop" (Bal 2016), the challenges and opportunities of the participation of academic institutions in the Anasagar Governance were discussed. Papers from diverse field were presented including the biodiversity, hydro-engineering, social engineering, ontology engineering, food security, climate change, social-ecological system approach, etc. The workshop concluded that studies on Anasagar must be compiled either at the local government level or at the university level to create a knowledge capital for information sharing that will facilitate public awareness and public participation besides helping in the decision-making (Bal 2016). The other nongovernment organizations also seem to have lesser participation in the Anasagar Governance, for example, the *Dargah* Committee (of the *Dargah Sharif*), appointed by the Government, takes care of the maintenance of the shrine and runs various charitable institutions like dispensaries, and guest houses for the pilgrims however do not participate in the lake-related activities.

For above three categories of the stakeholders, city and city infrastructure management is a "They" thing that is the local government's responsibility. The genesis of this perception among the stakeholders and local people comes from the city planning, management, and maintenance processes. Though the participation is considered in principle, their participation in the city planning (like the Anasagar restoration report/Master Development Plan (Draft) – Ajmer Region) is tokenism (Arnstein 1969) and in the implementation is almost negligible. In the urbanization process, the lake governance process (across the country) has changed into governmental when the government became the provider of the lake services and local people merely the user of the lake function. That distanced the local people from the governance process (in most city management) and brought the notion among them that "it is the government's job to maintain the lake." Amidst this, there are a number of indirect/direct peoples' initiatives that address, collaborate, and resist the current lake governance approach in their own way such as:

- The environment activism and awareness groups like the Rajasthan Samgrah Kalyan Sansthan are involved in women and biodiversity issues like birds.
- The print media like *Rajasthan Patrika* with its Ajmer edition are regularly raising issues of local people pertaining to Anasagar.
- Several health studies are carried out by the health inspectors and practitioners related to the impact of polluted water of the lake.
- The alternate healing group working on awareness of the people toward the relation between clean water and mind.
- The United Ajmer group brings local political parties together to work toward the physical improvement of the lake surroundings and its inlet drainage channels.

Local peoples' participation in lake governance is manifested through various stakeholder organizations discussed above. Besides, there is the urban community in general that is the user of the lake functions. Their use pattern (Fig. 9.9) influences the governance approach (Van Ast et al. 2013). Like other cities, the urban community of Ajmer is large and heterogeneous. The urban community belonging to different socioeconomic background extracts diverse option values. The urban community is divided into the resident local population and the floating pilgrim/tourist population. The local people perceive Anasagar as a wastewater reservoir, and the lake shores are the main public open spaces in the city for recreation and ritualistic activities besides groundwater recharging. The pilgrims (and tourists) perceive Anasagar as holy. Many of them take a dip in the lake and carry its water

Fig. 9.9 The urban occupations with Anasagar (Source: the author)

back homes considering it as "sacred water." There is a common belief among them that the water has healing properties for several kinds of illness and that taking a dip in the lake shall also purify the sins and fulfill ones' wishes. Both local residents and pilgrims acknowledge the aesthetic and ecosystem value of the lake. Both are contributing to the lake pollution directly (solid waste disposal) and indirectly (wastewater discharge). They can be referred as the stated and preferred value. Such values are reflected in the governance criteria that the urban community and the organizations also use to assess the improvements in the lake.

Second is the funding. ADA officials have confirmed the receipt of a sum of around 150+ million INR (as the allocated 70% of the total NLCP restoration budget from the Directorate through a state government organization) in the Board as of March 2014. The NLCP project ended in March 2014 including the grace period of 1 year. Since 2013, the NLCP and National Wetland Conservation Program are merged into one policy, namely, National Plan for Conservation of Aquatic Ecosystems (NPCA). The role of NCPA in Anasagar especially the funding is unclear. The ADA officials have funds remaining from the total fund received and are planning to utilize it for the other planned restoration activities in the grace period provided within the NLCP. The irony is that if the Board does not utilize the fund received within the grace period, it is subject to return to the Ministry of Environment and Forest, the status of which is unknown. In addition, the 30% fund shared between the state and local government is in a deadlock situation. The state government's concern is that the 20% fund shall be disbursed once the Board confirms the availability of 10% fund for the O&M, whereas the local government's challenge is that they need the 20% fund to invest and generate the 10% desired fund. The funding situation has slowed the restoration process. Funding is further challenged by the legal battle on property rights between the local people and the local government as the land parcels in the submergence area of the lake are to be compensated from this fund.

Third is the restoration on ground. The Board's focus on the reduction of pollution and soil erosion, regeneration and development of lake shore and catchment, and socioeconomic development of the people dependent on the lake is laid out in a Anasagar restoration report on "Lake Rejuvenation Project Anasagar, Ajmer," prepared by the IL&FS Ecosmart Limited and DHI India Water & Environment (Private) Ltd. (ADA 2007). The Anasagar restoration report highlighted the lake pollution and remedial measures besides shared responsibility of participation in the different lake activities by different stakeholders including the local people (IL&FS 2007). Sharma et al. (2008) analyzed the Anasagar restoration report and presented an integrated lake restoration model of Anasagar (in TAAL 2008) that included catchment area, institutional sustainability, and water as the key characteristics (Fig. 9.10).

Alongside NLCP, the Jawaharlal Nehru Urban Renewal Mission (JnNURM[11]) plays a crucial role in Anasagar restoration. It is a city infrastructure development program started in 2007 and involves physical integration of the drainage infrastructures such as rainwater and wastewater drainage and lake restoration. It involved preparation of a City Development Plan (CDP) by the ADA and AMC as a vision

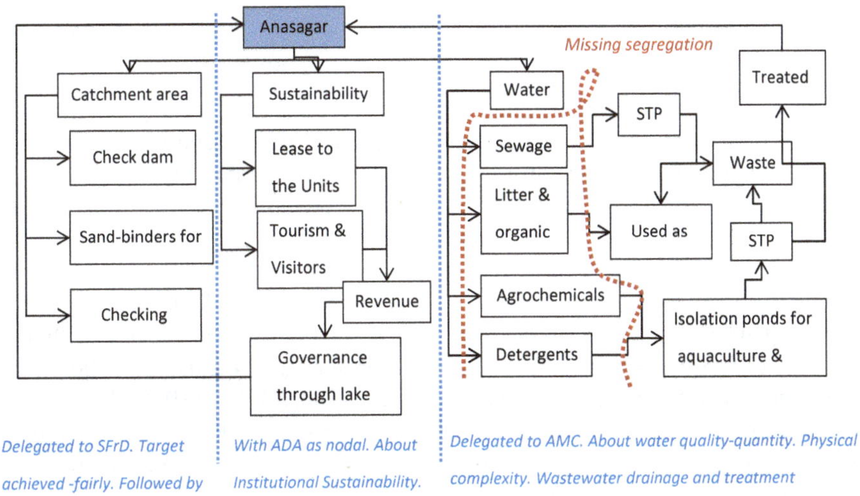

Fig. 9.10 Assessment of the integrated restoration plan for Anasagar (Source: Adapted from Sharma et al. 2009)

[11] The JnNURM was initiated in 2004 by the Ministry of Urban Development (MoUD) with the Ministry of Urban Environment and Poverty Alleviation (MoUEPA). The scheme is amended to "AMRUT" since 2014.

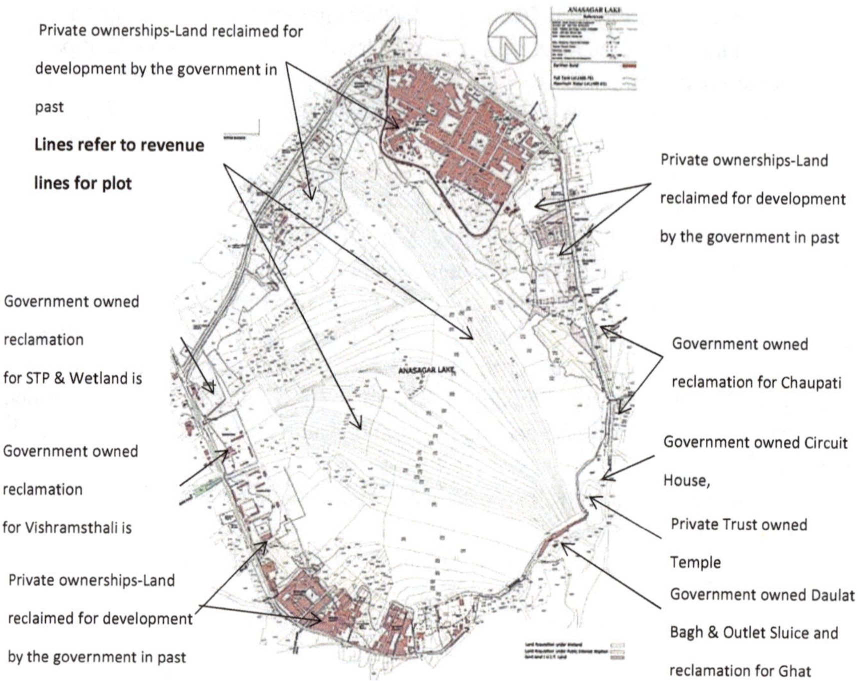

Fig. 9.11 Ownership of the land parcels in and around Anasagar (Source: Adapted from ADA, 2007)

document for the infrastructure development of the city. The funding structure and notion of participation and delegation of activities are similar to the NLCP.

The property right system of the land parcel encompassing the lake weakens the interaction between the local people and the local government. Anasagar is atypical lake in India where the property rights of the land encompassing the lake area belong to some thousands of landowners including local people and local government organizations unlike most lakes that are the national property with the Ministry of Environment and Forest locally represented by the district collector as guardian as per the Forest (Conservation) Act 1988 (MoEF 1990). Figure 9.11 shows the ownership pattern of the land parcels in and around Anasagar. Groups of local people holding property rights on the land parcel within the lake submergence area have come together for legal battles with the local government regarding the water level of the lake and the submergence of their lands. The reason of the tension is about the amount/logic of the solution offered by the Board that is not agreed upon by the local landowners. As an alternate to reclaim the land from the lake submergence, land reclamation is in progress at different parts of lake by the local people holding lands there since they perceive the land as lake shore (Fig. 9.12). This is also a case otherwise to the lake shores that are vulnerable to developments (Bal 2015a, b), and this is how Anasagar has lost its full tank area in the past.

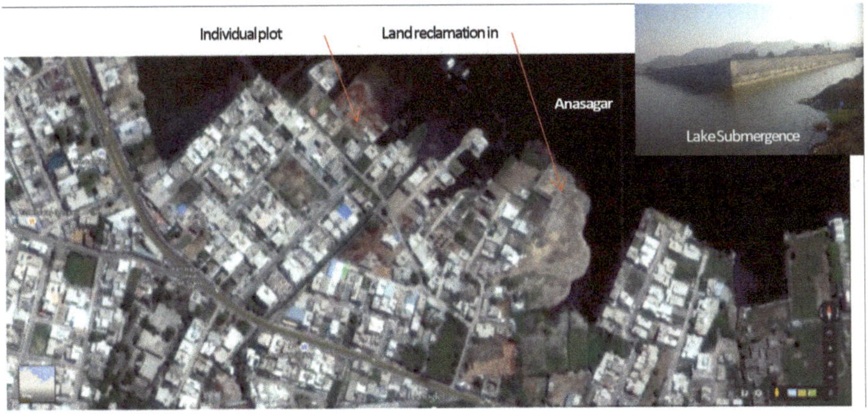

Fig. 9.12 Reclamation at the Lake Shore (Map Source: Google maps, accessed 10 May 2015)

6 Discussion and Conclusion

The paper discusses the various contextual (physical, institutional) and community characteristics that influence governance of Anasagar. The key contextual and institutional characteristics influencing the Anasagar governance and sustainability are the water level in the lake and the associated community issues pertaining to flood and the institutional issues pertaining to the property rights of the land parcels encompassing the lake. Besides, several other interesting results are found that shall shape the paradigm of interactive water governance in India with respect to urban lakes. The findings shall act as a feedback to the condition of Anasagar after recent interventions (like the NLCP and JnNURM) in terms of the physical condition and the pattern of interaction in the lake governance process. New issues and values have emerged that together will guide further studies on Anasagar. Governance is a cyclic: the value generated by the lake drives the governance processes and that in return changes the value of the lake over time. It is difficult to capture the exact time of change in value; however, the time of governance process is attempted here.

The framework used provides a guided understanding of the stakeholder participation alongside the other ILBM characteristics that facilitate/challenge interactive governance toward lake sustainability. The lake governance involves complex problems and processes built through the several simultaneous interactions between the different characteristics of the lake system. Therefore, though normatively, interactive governance is a way toward sustainability; in reality it leads to some clumsy solutions that have endemic contextual and institutional conflicts (of interests/priorities) leading to pluralism of policy approaches where participation, communication, information, finance, and technology still remain crucial for plausible selection of action that may/may not work in the benefit of the lake at a given point in time or in the long run. Thus, the contextual and institutional issues of Anasagar prior to the recent interventions continue, for example, the wastewater inflow, the property rights conflict, and the lack of knowledge capital.

Governing a lake system involves primarily land use planning and development of the infrastructure in the catchment that has several aspects to address bringing in the complexity. Finding the optimal solution that fosters urban development as well as conserves the lake ecosystem is a continued challenge. The land use planning rests over the dilemma of lake redevelopment/development and lake conservation approaches. The priority of lake development versus lake conservation requires different planning approach and that plays in the minds of the lake planners and politicians (Bal 2015a, b). A good example is the condition of the lake shore. The lake shore is the most vulnerable part of a lake system, and here the land-water nexus is evident linking it to the conflict of property rights and land reclamation.

The perception that local government apply to register improvements in the lake conditions is year-round availability of water in the lake, flourishing recreational activities, economic opportunities, urban community's acceptance of the restoration approach, etc. The notion of improvement is temporal given that over the last few decades, Anasagar is a cesspool of wastewater, portions of the lake are reclaimed for development, and there still remain some unattended lake shores. There is now year-round water and increased water level in the lake compared to the prior shrinking lake area and water level in the last few decades. Besides submergence of developments at the lake shore, there are regular flooding and backwater flow in the surrounding developments. The integrated drainage system and the associated wastewater treatments plants are the urgent needs for controlling lake pollution. The local people perceive the current state of water (level) to be excess than desired by them and that causes health issues with rise in water borne diseases and economic losses with loss of land in lake submergence and loss of farming in the dry season. If the improvements are compared with historic uses when the lake was used for domestic and irrigation purposes, then the present health of the lake is questionable. Furthermore, the notion of improvement is related with a time in the future in the context of delineation of the lake boundary, improved strategy to ensure protection of the lake ecosystem, information sharing and communication to enhance participation of the local people and other marginalized institutions like the academics and businesses, and improved arrangement of technical, financial and administrative resources, etc. Ajmer is likely to be included in the national program of "Smart Cities" development and that is expected to facilitate the improvements with respect to Anasagar.

While the literature on shift from government to governance (Loorbach 2007; Kickert 1997; Kooiman 1993) in water management is on the rise, the ground realities are somewhat different in the case of urban lake management in India. The water management is still by far a government thing working from/at multiple levels and scales and not as interactive as it should be.

The interactive governance in Anasagar thus is similar to what the Ostroms (Ostrom and Ostrom 1977) call as polycentric governance in which authorities (organizations here) from overlapping jurisdictions (or centers of authority) interact to determine the conditions under which these authorities, as well as the communities subject to these jurisdictional units, are authorized to act as well as the constraints put upon their activities for public purposes (McGinnis 2011). While the community as group is indirectly involved in the governance through various activities, however the fact that the government officials are from the local community must be considered influencing the decision-making process.

The activities carried out by the Board through the NLCP are a manifestation of the interactive governance process. The process tempts to link the interactive process to the logic of the collectivity (Olson 1965). At the face value, the participating organizations made a collective choice to improve Anasagar, who otherwise worked independently and in uncoordinated manner in the past that had devoid them in producing cumulative positive outcomes. At the core, there are specific social and ecological events and disturbances (Schoon and Cox 2012) that left the organizations with no choice than to engage in the interactive process toward Anasagar restoration. The NLCP and the JnNURM and the water woes linked to the property rights are some key events and disturbances that facilitated the interactive governance. These events and disturbance also remain crucial in sustaining as well as disrupting the interactive governance over time. There is interdependency of actions between the organizations. On the one hand, it has turned useful in monitoring the actions of different organizations and for accountability; on the other hand, there are delegations of responsibilities, lobbying activities and interest groups formations between the officials of the different organizations, in order to pursue or not pursue certain activities (Bal 2015a, b).

The participation of the local people in the governance process is challenged by the tension between them and the local government with regard to property rights, water level leading to excess water problems, tokenism in governance process, poor communication of the Board about the restoration process, and visible slow improvement in the lake and the surrounding developments. For the governance process to be more inclusive of local people, the start point is the knowledge integration of the data and activities carried out by the different organizations in order to do integrated reporting (monitoring and evaluation) on Anasagar and then disseminate the information to the local people. Developing a lake management and interpretation body comprising of local stakeholders from various backgrounds may be a good idea at this point. This will engage local people in understanding the issues of Anasagar besides either enjoying or worrying its presence and thereby is likely to enhance participation in the governance process. This may also change the perception of the local government and the people about government-induced governance and may make the governance more resilient to social and ecological changes.

Acknowledgment I thank the reviewers for the insights in refining the direction and presentation of the paper, the IDE-JETRO for the Research Grant (2016–17) for the project 'Interactive Approaches to Water Governance: Case Studies in Asia, in particular thanks to Kenji Otsuka for leading the project, the Research Institute for Humanity and Nature, RIHN, Kyoto for the Research Fellowship (2014–15) to carry out an action research on Anasagar. My sincere thanks extends to the local organizations and the people of Ajmer for being instrumental during the fieldwork and the Maharishi Dayanand Sarasvati University Ajmer for co-hosting the Workshop 'Water and Wetland' for knowledge exchange on Anasagar.

Appendices

Appendix 1: Summary of the Issues of Anasagar Published in Newspaper Article of October 2014

The title of the article is Development in the Lake or Lake in the Development. The other captions are (1) for the last 3 years, the developments are submerged in the lake water. The district administration is unable to take remedial decisions; (2) submerged development and emotional outburst; and (3) until when the building is going to remain submerged. Submerged developments in the lake since 2010 include the government buildings like the Vishram Sthali, the secondary treatment plant, and the constructed wetland besides several private residential and commercial buildings. Some of the highlighted issues in the article are the water level issue continues; the relocation is not easy; and the image of Ajmer is at risk.

अजमेर, सोमवार, 06.10.2014 rajasthanpatrika.com अजमेर पत्रिका patrika.com राजस्थान पत्रिका

आनासागर › तीन साल से झील के पानी से घिरी हैं बस्तियां, जल्दी निर्णय नहीं कर पा रहा जिला प्रशासन

झील में बस्तियां या बस्तियों में झील?

डूबती बस्तियां, छलकता दर्द

मांग रहे बाजार भाव का मुआवजा, भूमि के बदले भूमि की मांग

झील का स्तर बरकरार

अजमेर की बिगड़ती छवि

पुनर्वास आसान नहीं

कब आएगी डी-वीडिंग मशीन?

कब तक डूबी रहेगी विश्राम स्थली?

क्या यहां टी-पाइंट बनेगा?

यूं छलकी पीड़ा

Source: Ajmer Patrika, Oct 06, 2014

Appendix 2: Different Programs Objectives and Activities

Objectives	**Activities**
NLCP	
1. In situ measures of lake cleaning such as de-silting, de-weeding, bioremediation, aeration, bio-manipulation, nutrient reduction through constructed wetland approach or other successfully tested eco-technologies, etc.	Physical
2. Catchment area treatment includes wastewater-rainwater treatment, afforestation, silt traps, check dams, etc.	
3. Strengthening/protecting the embankment, lake fencing, etc.	
4. Lake front/shoreline development including public interface	
5. Prevention of pollution from nonpoint sources by providing low-cost sanitation	
6. Public awareness and public participation	Social
7. Capacity building, training, and research in the area of lake conservation	
8. Any other activity depending upon location-specific requirements – unclear	Challenge
Excluded but crucial: solid waste management and provision of Dhobi Ghats	
↑Source: Adapted from MoEF (2008)	
JnNURM	
1. Attention to an integrated approach toward infrastructure development	Physical
2. Ensuring adequate funds to meet the deficiencies of urban infrastructural services	
3. Focus on water supply and sanitation, solid waste management, road network, urban transport, and redevelopment of old city areas	
4. Basic infrastructure services to the urban poor	
3 and 4 are indirectly linked to the restoration of the urban lakes.	
5. Establish linkages between asset creation and asset management through reforms for long-term project sustainability of infrastructure services	Social
↑Source: MoUD and MoUEPA (2007)	

References

ADA. 2007. *Detail Project Report on Anasagar Lake Project*. Ajmer: Ajmer Development Authority.

———. 2013. *Master Development Plan (Draft)-Ajmer The Region in Hindi*. Ajmer: Ajmer Development Authority.

Agarwal, A., and S. Narain. 2001. *Dying Wisdom*. New Delhi: Centre for Science and Environment, CSE.

Agarwal, A., S. Narain, and I. Khurana. 2001. *Making Water Everybody's Business: Practice and Policy of Water Harvesting*. Delhi: Centre for Science and Environment.

Agrawal, Y.C. n.d. *Catchment Areas and Their Restoration*. Jaipur: Water Resources Department, Government of Rajasthan State of India.

AMC and PMB. 2006. *City Development Plan For Ajmer and Pushkar*. Ajmer: Ajmer Municipal Corporation and Pushkar Municipal Board.

Arnstein, S.R. 1969. A ladder of citizen participation. *Journal of the American Institute of planners* 35 (4): 216–224.

Bal, M. 2015a. *Social-ecological System Framework. Understanding Urban Lake Governance and Sustainability in India*. In *PhD thesis published at the Erasmus University Rotterdam* http://repub.eur.nl/pub/77643.

———. 2015b. *A Social-Ecological system approach to analyse Anasagar Governance*. Report submitted to the Research Institute for Humanity and Nature, RIHN Kyoto.

———. 2016. *Proceedings of the Workshop 'Water and Wetland: Looking For Smart Solutions'*. MDS University Ajmer on 19 Mar 2016. Paper presentation on 'Analysing Anasagar from a Social-Ecological system approach'.

Berkes, F., J. Colding, and C. Folke. 2003. *Navigating Social-ecological Systems: Building Resilience for Complexity and Change*. New York: Cambridge University Press.

Edelenbos, J. 2005. Institutional Implications of Interactive Governance: Insights from Dutch Practice. *Governance: An International Journal of Policy, Administration, and Institutions* 18 (1): 111–134.

Edelenbos, J., and I.F. Van Meerkerk. 2017. Normative Theory. In *Handbook on Theories of Governance*, ed. J. Torfing and C. Ansell. Cheltenham: Edward Elgar Publishers.

GWP-TAC. 2000. *Integrated Water Resource Management Tool Box Version 2. Global Water Partnership*.

IL&FS. 2007. *Detail Project Report on Anasagar Lake Project*.

ILEC. 2007. *Integrated Lake Basin Management: An Introduction*. Kusatsu: International Lake Environment Committee Foundation http://www.ilec.or.jp/en/.

JnNURM. 2005. *Internal Earmarking of Funds for Services to Urban Poor, ULB Level Reform. Mandatory Premier*. Jawaharlal Nehru National Urban Renewal Mission, Government of India.

Kickert, W.J.M. 1997. Public governance in the Netherlands: An alternative to Anglo-American 'manegerialism'. *Public Administration* 75 (4): 731–752.

Kiser, L.L., and E. Ostrom. 1982. The Three Worlds of Action: A Metatheoretical Synthesis of Institutional Approaches. In *Strategies of Political Inquiry*, ed. E. Ostrom, 179–222. Beverly Hills: Sage Reprinted in McGinnis, 2000.

Kooiman, J. 1993. Findings, Speculations and Recommendations. In *Modern Governance: New Government – Society Interactions*, ed. J. Kooiman, 249–262. London: Sage.

Kooiman, J. 2003. *Governing as Governance. Part III*. London: Sage.

Lintner, S.F. 2007. *Opportunities and Challenges in Integrated Lake Basin Management*. Shiga: ILEC, International Lake Environment Committee Foundation.

Loorbach, D. 2007. *Transition Management: New Mode of Governance for Sustainable Development*. Utrecht: International Books.

McGinnis, M. 2011. An Introduction to IAD and the Language of the Ostrom Workshop: A Simple Guide to a Complex Framework. *The Policy Studies Journal* 39 (1): 169–177.

Ministry of Urban Employment and Poverty Alleviation (MoUEaPA) and Ministry of Urban Development (MoUD). 2007. *Jawaharlal Nehru National Urban Renewal Mission (JnNURM): Over view. MoUEaPA and MoUD*. New Delhi: Government of India.

MoEF. 1990. *Wetlands of India -A Directory*. New Delhi: Ministry of Environment and Forests. Government of India.

———. 2004–05. *Annual Report. Conservation of Water Bodies*. Ministry of Environment and Forests, Government of India, New Delhi.

———. 2008. *Guidelines for National Lake Conservation Plan*. New Delhi: National River Conservation Directorate. MoEF, GoI.

Narayanan, N.C., and J.P. Venot. 2009. Drivers of Change in Fragile Environments: Challenges to Governance in Indian Wetlands. *Natural Resources Forum* 33 (4): 320–333 Blackwell Publishing Ltd.

Olson, M. 1965. *The Logic of Collective Action: Public Goods and the Theory of Groups*. Cambridge: Harvard University Press.

Ostrom, E. 1990. *Governing the Commons: The Evolution of Institutions for Collective Action*. New York: Cambridge University Press.

———. 2005. *Understanding Institutional Diversity*. Princeton: Princeton University Press.

———. 2009. A General Framework for Analyzing Sustainability of Social-Ecological Systems. *Science* 325 (5939): 419–422.

Ostrom, V., and E. Ostrom. 1977. A Theory for Institutional Analysis of Common Pool Problems. In *Managing the Commons*, 157–172. San Francisco: W.H. Freeman.

Pandey, D.N., B. Gopal, and K.C. Sharma. n.d. *Compiled a List of Publications on Anasagar in Their Report on 'Evidence-Based Holistic Restoration of Lake Anasagar' Submitted to the High Court of Rajasthan.*

Peters, B.G., and J. Pierre. (draft). 2012. *Interactive Governance.* Pittsburgh: Department of Political Science, University of Pittsburgh.

Pierre, J., ed. 1998. *Partnerships in Urban Governance: European and American Experience.* Basingstoke: Macmillan.

Putnam, R. 2000. *Bowling Alone: The Collapse and Revival of American Community.* New York: Simon & Schuster.

Sarda, H.B. 1911. *AFMER: Historical and Descriptive.* Ajmer: Scottish Mission Industries Company Limited.

Schoon, M.L., and M.E. Cox. 2012. Understanding Disturbances and Responses in Social-ecological Systems. *Society & Natural Resources* 25 (2): 141–155.

Sharma, K.C., P.D. Charan, and M. Nag. 2009. *An Integrated Restoration Plan for Lake Anasagar: Athreatened Water Body of Ajmer, Rajasthan, India.* In Proceedings of 13th World Lake Conference, 1–5 November, Wuhan, China 2009.

Sharma, K.P. Sharma, S. Sharma, S. Sharma, P.K. Swami, R.C. Singh, P.K. & Rathore, G.S., 2008. Mansagar Lake: Past, Present and Future. In *TAAL 2007. Jaipur: Proceedings of the 12th World Lake Conference*, ed. M. Sengupta and R. Dalwani, Oct 29–Nov 02, 2007.

Sørensen, E., and J. Torfing, eds. 2007. *Theories of democratic network governance.* Cheltenham: Edward Elgar.

TAAL. 2008. Jaipur Declaration. In *TAAL 2007. Jaipur: Proceedings of the 12th World Lake Conference*, ed. M. Sengupta and R. Dalwani, Oct 29–Nov 02, 2007.

Torfing, J., B.G. Peters, J. Pierre, and E. Sørensen. 2012. *Interactive Governance: Advancing the Paradigm.* Oxford: Oxford University Press.

Van Ast, J.A., J.J. Bouma, and M. Bal. 2013. Values Connecting Societies and Water Systems. In *Water Governance as Connective Capacity*, ed. J. Edelenbos, N. Bressers, and P. Scholten, 249–266. Surrey: Ashgate Publishing Limited.

Mansee Bhargava is an independent entrepreneur, researcher, and educator at Environmental Design Consultants, Ahmedabad. Her practice, research, and teaching focus are on large-scale residential, institutional, recreational, and industrial developments with ecological inclination. She believes in sustainability and livability aspects of design, development, planning, management, and governance of human settlements and adheres to them in diverse engagements. In the paper on Ana Sagar, she highlights the interlinking and complex nature of social-ecological systems particularly urban lake system of India.

FSC
www.fsc.org
MIX
Papier | Fördert
gute Waldnutzung
FSC® C083411

Kolibri 360 GmbH
Ferdinand-Jühlke-Straße 7
99095 Erfurt, Deutschland
produktsicherheit@kolibri360.de